£22.50

OXFORD STUDIES

IN

NUCLEAR PHYSICS

GENERAL EDITOR

P. E. HODGSON

DIFFRACTIVE PROCESSES IN NUCLEAR PHYSICS

W. E. FRAHN

Formerly Professor of Physics, University of Cape Town

CLARENDON PRESS · OXFORD
1985

Oxford University Press, Walton Street, Oxford OX2 6DP
Oxford New York Toronto
Delhi Bombay Calcutta Madras Karachi
Kuala Lumpur Singapore Hong Kong Tokyo
Nairobi Dar es Salaam Cape Town
Melbourne Auckland
and associated companies in
Beirut Berlin Ibadan Mexico City Nicosia

Oxford is a trade mark of Oxford University Press

Published in the United States
by Oxford University Press, New York

British Library Cataloguing in Publication Data
Frahn, W. E.
Diffractive processes in nuclear physics.—
(Oxford studies in nuclear physics)
1. Diffraction
I. Title
539.2 QC415
ISBN 0-19-851512-X

Library of Congress Cataloging in Publication Data
Frahn, W. E.
Diffractive processes in nuclear physics.
(Oxford studies in nuclear physics)
Includes bibliographies and indexes.
1. Scattering (Physics) 2. Diffraction.
I. Title. II. Series.
QC794.6.S3F73 1985 539.7′54 84-19966
ISBN 0-19-851512-X

Set and printed in the United Kingdom by
The Universities Press, (Belfast) Ltd, Northern Ireland

AUTHOR'S PREFACE

The purpose of writing this book is primarily a personal one. Throughout my scientific life I have been attracted and enchanted by the beauty and variety of quantal interference phenomena, and I am a 'nuclear physicist' only inasmuch as these manifest themselves with particular clarity in nuclear processes. Because the works of those who have a similar outlook, and my own contributions, are dispersed in numerous original papers, I reacted positively to Peter Hodgson's suggestion, in 1973, to write a coherent account of diffraction scattering in nuclear physics. At that time, however, the theoretical development was still rudimentary and most of the high-precision experimental work—especially on heavy ion collisions—had yet to be done. Only at the present time do I feel that our understanding has reached a certain measure of completeness, and that a book of this nature might be called for. If some of my peers think likewise . . . .

How does one go about writing such a book? Of course, I want to highlight what I consider to be the main development. This is presented in Chapter 6, which is the longest part of this book and, frankly, my emphasis lies there. A general motivation has to be provided; this is attempted in Chapter 1. Basic concepts are explained in Chapters 2 and 3. The evolution that has led to the more recent developments is traced in Chapter 4; although most of it is now obsolete, it had a seminal influence and is recounted here for historical continuity. Other, alternative or conflicting, approaches are sketched in Chapter 5; if less justice than they deserve is given by my presentation, I apologize. Inelastic scattering, transfer reactions and coupled-channels theories are described in Chapters 7, 8 and 9.

I thank all my collaborators and students, and also many colleagues, for discussions, the typist, the Press, and colleagues and publishers for permission to reproduce figures.

Cape Town W. E. F.
1982

EDITOR'S PREFACE

In January 1973 I invited Professor Frahn to write a book on *Diffraction scattering in nuclear physics*, and he immediately replied that he would like to do so because 'this seems to be an opportune moment now that the systematic features of nuclear diffractive scattering have become clear and the subject has reached a certain stage of completion'. By March he had drafted the outline of the chapters, and this was accepted by the Oxford University Press.

Over the following years he worked on the book whenever he could, but he frequently found that his original plans had to be modified in the light of the latest developments. He worked intensively with several collaborators and together they applied the methods of diffraction analysis to elastic scattering, inelastic scattering, and nucleon transfer reactions. Whenever he saw that a new understanding could be achieved he laid the book aside and concentrated on breaking the new ground and applying the new formalism to the analysis of experimental data. This fruitful process of research and writing was tragically terminated by his death on 19 April 1982.

Among his papers there was found sufficient material to form the major part of his proposed book. It is not as complete as he would have wished, and certainly he would have further amplified and developed many of the chapters. It is necessary to bear these circumstances in mind when reading this book.

Over the years of preparation, Professor Frahn frequently sent reports on his progress. As these provide both a summary of the development of diffraction theory and the essential basis for understanding the plan of his book the main points are reproduced here.

In December 1973 he wrote:

> I have been working on the subject matter of the book continuously since our meeting in June 1973. However, most of this work was concerned with major new developments that have taken place during this time as a result of the unprecedented progress in heavy ion physics, which has become the most active frontier of nuclear research. As you know, heavy ion reactions provide the most important examples of nuclear diffraction scattering. For instance, our early predictions of diffractive structure of heavy ion transfer cross sections were first

confirmed experimentally just over a year ago and the evidence has since been strengthened by a wealth of new data from many laboratories. Concurrently there have been new developments of the theory, such as semiquantal theories with complex trajectories, applications of Regge pole and Glauber theory, rainbow scattering, etc. I myself have developed a new unified quantal formulation of diffraction scattering which generalizes the earlier work and incorporates, or closely relates to, the other recent theoretical approaches. This theory has attained its final form only a few weeks ago and numerical evaluation is still in progress in Munich, Berlin and Copenhagen. The field is clearly of high topical interest and has been the subject of several international conferences this year, the latest one at Heidelberg in October in which I have taken an active part. Had I written the book on the basis of our knowledge as of June 1973 it would have been largely out of date by now.

However, I feel that now we have reached a stage of consolidation and I can start with the actual writing. The overall plan of the book is unchanged and I have assembled most of the material for each chapter. Thus I still expect to be able to meet the original deadline and have the typescript ready by the end of 1975. In case I shall have to ask for a little extra time it would not be due to lack of effort on my part but to the circumstances mentioned above. I have a keen interest in getting the book ready as soon as possible because I am sure the need for it has greatly increased since the time it was first proposed.

In August 1975 he reported that:

I have made substantial progress in the theory of heavy-ion collisions of which I want to give you a brief outline.

1. *Elastic scattering*

By the end of last year I had developed a new systematic method for closed-form evaluation of the quantal scattering amplitude; this is described in the revised version of my 1973 Trieste lectures which will soon be published by IAEA. During my stay in Berlin in March and April this year the theory has been extended and put on a firm mathematical foundation in collaboration with Dieter Gross. It has now attained its definite final form and I am at present writing a comprehensive paper (jointly with D. Gross) entitled "Closed-form evaluation of the amplitude for elastic heavy-ion scattering". I am convinced the physics and mathematics of the subject is now completely understood.

2. *Inelastic scattering*

While in Munich in June–July last year, I had given a closed formulation of inelastic heavy-ion scattering including nuclear and Coulomb excitation. This was subsequently reformulated by means of a similar systematic method as for elastic scattering. It is now also in its final form. The resulting closed expressions for both elastic and inelastic scattering have been evaluated numerically by K. E. Rehm in Munich, who has written a handy computer programme for analysis of experimental data. This is now being applied to all available heavy-ion data with most satisfactory results. As soon as this project is completed, the theory and the results of the comprehensive analysis will be reported in a joint paper with K. E. Rehm.

3. *Transfer reactions*

A similar closed evaluation of the DWBA amplitude for heavy-ion transfer reactions has been presented in the Trieste lectures and in my talk at the Heidelberg symposium last October. This is now being extended and refined to include recoil and coupled channels contributions.

4. *Polarization*

During my recent stay in Frankfurt and Heidelberg I have developed closed expressions for the vector and tensor polarizations in elastic scattering of spin-$\frac{1}{2}$ and spin-1 heavy ions. This work was motivated by the very interesting first results obtained by the group of D. Fick at the MPI Heidelberg with their new method of producing polarized ^{6}Li beams; they expect to get a wealth of polarization data with this and other alkali (and possibly also halogen) ions when their source is in full operation by the end of this year.

5. *Relation to other approaches*

As you know, the main alternative (or rival) formulations of heavy-ion collisions are the semiclassical, complex trajectory method of Knoll and Schaeffer at Saclay, and the Regge pole descriptions of Fuller and McVoy, Marty and Rowley, and others. While the relation of the latter approach to my own is pretty clear and non-controversial, my recent discussions at Saclay with Jörn Knoll and Richard Schaeffer (in which John Blair took part) have been concerned with considerable differences of opinion regarding the physical interpretation of their results, particularly with respect to diffraction scattering. These differences have only partially been resolved, mainly because their work is by no means complete as far as application to data analysis is concerned. Personally I have a clear picture of what is going on, but we'll have to await their further results before full agreement may be reached.

I'm telling you all this because these developments obviously affect the writing of the proposed book on diffraction scattering, for which heavy-ion collisions are one of the most important areas of application. The subject has entered an entirely new phase since we first discussed the plan of the book in 1973, at a time when I felt it had reached "a certain stage of completion". This is still true as far as the basic concepts of diffraction scattering are concerned, but these have only now been based on a firm foundation such that what had earlier been little more than a heuristic form of description has now become a rigorous, quantitative part of quantal collision theory. Moreover, the range of nuclear diffraction scattering has been vastly extended by the wealth of heavy-ion data in the last few years. It is also only during this time that the semiclassical approaches, Regge pole methods, etc., have been developed. I have at various stages written several versions of practically all the chapters of the book, but had to revise them over and over again in the light of the development of my own work and that of others. The subject has been in such a state of flux that an attempt at consolidation at any earlier stage would have been premature. I am, however, confident that such an attempt can now be made. This will, I think, become clear when the two papers mentioned under 1 and 2 above are completed, at which I am working flat out but will take at least to end of this year.

This means, unfortunately, that I have to ask for an extension of the delivery date for the manuscript. While I'm sorry about the length of this letter, I would like you to see why this request has become necessary.

In June 1976 further progress was reported:

During the past months I've been working very hard to finish my work on inelastic heavy ion scattering. One paper describing the theory is now completed and I shall send you a copy within the next few days. This will have some surprises for many people, showing that wave-mechanical "fact" is quite different from semiclassical "fiction". In another long paper, written together with Ernst Rehm of the Technical University of Munich, we present the results of analyses, by means of my closed formalism, of a large number of elastic and inelastic heavy-ion data: angular distributions and excitation functions, for target and projectile excitation, for the M-substate populations, reorientation effect, etc. This will probably be finished soon after my return from the Gordon Conference on Nuclear Chemistry and a subsequent visit to Germany for which I leave at the end of this week. In addition we have sent off two letters reporting the highlights of our results—I wonder what reception they will get from the semiclassical fraternity!

An earlier, and even longer, paper with Dieter Gross of the Hahn-Meitner-Institute in Berlin, in which we develop the closed formalism for elastic heavy-ion scattering, has just been accepted by *Annals of Physics*. This is mainly of a mathematical nature; the resulting explicit expression for the elastic scattering amplitude is identical with that given in my latest Trieste lectures.

As you mention "diffraction theories", I should perhaps point out, and in fact emphasize, that the closed formalism is not another diffraction theory, but an explicit representation of the actual optical model and DWBA-amplitudes, which comprises all the leading terms of an *asymptotic expansion* with respect to the large parameters of heavy ion interaction. I believe that we now have for the first time a complete and detailed understanding of the physics that is implicitly contained in optical model and DWBA computations for heavy-ion scattering, and I hope you will find that the above-mentioned papers substantiate this belief.

Now that the theory has been developed I am at last in a position to go ahead with the book on "Diffraction Scattering in Nuclear Physics" which therefore will contain much new material.

Once again, however, new avenues of research opened up, and in January 1979 he wrote:

I ought to have kept you informed about the position regarding my text, only it is rather embarrassing to report again "sorry, not yet"! The reasons are the same as when I wrote to you last: the theory which will form a substantial part of the book is still not quite finished. However, most of the projects I mentioned in my last letter are now completed and the papers thereon published. The main topic still outstanding is transfer reactions for which the formalism has however been developed. I am collaborating with Ernst Rehm in Munich, who has written a programme for it, on the analysis of relevant data.

The first results we got during my visit there last October are very encouraging, and I hope we'll have a lot of analyses soon. Then I shall have all the material for a comprehensive presentation.

In a few days' time I shall leave again for a two-months' stay in Brazil, where I am to give an extensive course of lectures at the University of São Paulo. This will be a good opportunity to present a coherent account of my formalism and its applications, and thus a preparation for the monograph. I certainly have not lost the book out of sight, but I want to make it fairly complete in coverage. Its final form is now taking shape, and when I have put it together it will be as early and up-to-date as possible.

By February 1980 he was finally in a position to complete the book:

I have completed my papers, all of which are now or very soon being published, on the following topics:

(1) Large-angle heavy-ion scattering, elastic and inelastic (*Nucl. Phys. A*),
(2) Polarization in light and heavy ion scattering, elastic (*Ann. Phys. N.Y.*) and inelastic (*S. Afr. J. Phys.*),
(3) Heavy-ion transfer reactions (*Phys. Rev. C*).

During my stay in Brazil last year, my closed formalism has been extended to include *coupled-channels* effects and several papers on this, jointly with M. S. Hussein, have been completed. There is only one paper, on channel coupling contributions to inelastic heavy-ion scattering, which I am still writing up. With this, my programme of a closed-form description of elastic and quasielastic heavy-ion collisions is essentially complete. The theory is now being applied in extensive data analyses by several people, in Munich, São Paulo and Rio, and I hope we shall have lots of results soon.

Thus I have now reached the stage where the book can finally be written with a complete and coherent account of the developments mentioned above; it couldn't be done sooner. The contents will still be according to the outline I gave you some time ago, but the formulation can now be made much more comprehensive, concise and elegant. As long as I was still working on the development of the basic theory, the book was something of an embarrassment, but now I am keenly interested myself in putting all the different pieces together in a coherent story. I am confident that most of it will be ready by the end of the year.

I apologize for having taken so much longer than originally envisaged, but the extent of the groundwork that had to be done—on which I have worked flat out during the years since we first planned the book—could not have been foreseen. I believe, however, that the results have been worth the effort.

By April 1980 he had made good progress:

To show you that I am now really at the job of writing my book, I am sending you, under separate cover, a draft of the first chapter I have written so far. Herewith I enclose what I consider my definite proposal for the list of contents. First you will note that I have changed the title from "Diffraction Scattering" to "Diffractive Processes in Nuclear Physics", because the latter defines more accurately the scope of the subject; I hope you are agreeable to this.

Let me now give you a brief résumé of what I intend putting in the various chapters.

Chapter 1 is a general introduction to diffraction phenomena in nuclear physics, how they originate in the general concept of Quantum Diffraction, what is the aim and scope of diffraction theories, and what's their use in applications.

Chapter 2 collects basic material of scattering and reaction theory, only so far as it is needed in further developments: S-matrix, Coulomb scattering, spin formalism and polarization, identical particles, unitarity and absorption, integrated cross sections, optical theorem etc., for elastic scattering. For reactions, the basics of DWBA and coupled-channels theories.

Chapter 3 deals with complex potentials, absorption and continuity equation, general relations between S-matrix, wave functions and potentials. This is followed by a description of configuration space diffraction theory, i.e. the quantal analog of Fresnel–Kirchhoff diffraction for the Schrödinger equation, and its extensions to inelastic scattering and direct reactions.

Chapter 4 gives an account, in historical sequence, of the developments and results of earlier diffraction models: from Placzek and Bethe via Blair, Dar, de-Shalit, Inopin, Strutinsky and the many other diffractionalists, the early strong absorption model and its applications, to the simple Fresnel model and its generalization.

Chapter 5 discusses as briefly as possible how diffraction has been treated in the semiclassical theories, based on WKB and similar approaches: from Ford and Wheeler, Broglia and Winther, to complex trajectory formalisms, and the various "Regge pole" descriptions.

In *Chapter 6* I begin with a coherent account of what I call the Closed Formalism, from elastic scattering to coupled channels in Chapter 12. This Part II is a unified whole of analytic theory and is, as far as I am concerned, the centrepiece of the book around which the other parts are draped as introductions and extensions, respectively. Because I am most motivated in writing this part, I have started with it and it is of Chapter 6 that I am sending you my first draft. I shall continue drafting the other chapters of Part II, sending them to you piecemeal for your comments, and not begin on the other parts before this is completed.

Chapter 13 will give an account essentially of Glauber theory and its extensions beyond the eikonal approximation, such as our Fresnel expansions in potential and multiple scattering. Finally, in *Chapter 14* I present, mainly for completeness, a description of high-energy diffractive processes, in a somewhat extended version of what I have given in Section 6 of my paper on "Quantum Diffraction" in *Rivista*. Some formal developments that are necessary but would interrupt the flow of the main text will be collected in Appendices.

I would greatly appreciate your comments on this résumé and whether it meets with what you had in mind when you suggested I should write something like this.

Let me add a few remarks on the drafted Chapter 6. This is the longest and most 'formalistic' chapter of all, i.e. the others will be considerably shorter and have a lower formula density. But it forms the basis of the Closed Formalism and reflects the fact that elastic strong-absorption scattering has turned out to be in a sense the most subtle of the peripheral processes as far as the closed wave-mechanical description is concerned.

I must admit that I am rather pleased with the formulation as it now stands in Chapter 6. Firstly, the formalism for "neutral" scattering is new and has not been given in this generality anywhere else. Secondly, on redeveloping the theory for charged particle scattering I have found a way of treating the all-important Fourier transforms *to all orders;* it had bothered me for some time that in earlier formulations I had included them in first order only. Quite surprisingly, the result is as simple as the first-order formulae! It has all the right properties, and I consider the new expressions as *the* closed-form representation of the elastic scattering amplitude. I also hope you will like the simple qualitative interpretation of the formalism I have given in subsection 6.5.5 on "pictorial explanation". Of the figures I have appended only rough qualitative sketches indicating what I have in mind for illustration. I foresee that the preparation of new figures will give me the most trouble, because I have to rely largely on the help of others to do the required numerical calculations (I myself am hopeless with computers!).

The formalism in the subsequent chapters on inelastic scattering, transfer reactions, etc., will also be redeveloped in the light of what I have done for elastic scattering, but this will be fairly straightforward. This will make Part II a coherent whole which in this comprehensive form has not been presented before. You can imagine that I am now writing this out for my own pleasure as an exercise in aesthetics; if others will like it too I shall be pleased, but that's not my main concern.

In May 1980 he replied to my letter of encouragement:

I am very pleased that you like the summary of contents and the draft of Chapter 6, and I am very grateful for your good advice and suggestions. Please regard what I am sending you as first drafts; everything will be polished over again after I have completed the first write-up. I shall certainly heed your advice to indicate at the beginning of each section what the aim of it is. As regards the comparison of the theory with experimental data, this is what I most definitely intend to do. I didn't include any data analyses with the material I sent you because my main concern was to get the theoretical development in its new shape. I wanted to leave the data analyses to a later stage because this will give no more than practical problems of doing the computations and preparing the drawings—I know that the theory fits the data very well. In fact I also intend to show, for at least one typical example in each chapter, a comparison between the optical model or DWBA with the closed formalism, using the same S-matrix in both cases, to demonstrate the quantitative accuracy of the theory.

I have now completed the draft of Chapter 7 on elastic scattering. The theory for this I have re-developed *ab initio*, and I'm pleased to say that I found it possible again to sum the series over the Fourier transforms to all orders. Thus the new formulation represents a similar generalization of my previous (published) formalism as I gave in Chapter 6 for elastic scattering. Again, to my surprise, the structure of the final result is no more complex than the earlier form; in fact it is even simpler and more transparent conceptually. So both

Chapters 6 and 7 contain essentially new material; the rest of Part II will give a summary of results already described in the corresponding published papers, except for Chapter 10 which will also include a good deal of hitherto unpublished work.

Once again, however, the work on the book was interrupted, and in November 1981 he wrote, in what was to be his last letter:

I am very sorry for the delay in replying to your letter of 6 August. The work on my book has again been retarded because I had to write a review article on heavy-ion elastic scattering for Allan Bromley's project "Heavy Ion Science", which is more a summary of my particular approach to the subject (including very recent, unpublished work) rather than a pedagogical text. Also, again, a number of new projects have come up during and after my stay at the Hahn-Meitner Institute, Berlin, in mid-year, which are intriguing enough to call for intensive attention. For one, the old problem of anomalous large-angle scattering, in particular the structure of the 180° excitation function for $^{16}O+{}^{28}Si$ and similar systems, is very much with us. Following up a hunch I had in Berlin, I was able to calculate analytically the effect of angle-averaging implicit in the measured data, which our subsequent numerical calculations showed to be quite drastic, as described in the enclosed communication. This, to my mind, invalidates almost all previous theoretical work directed at the "wrong" data, and calls for a drastic revision of our ideas as to the physical nature of the large-angle enhancement processes. Then there are the new, good-resolution data on $^{12}C+{}^{12}C$ scattering at higher energies, showing intriguing low-l transparency effects about which we hope to be able to say something soon. Thus a number of important questions remain to be clarified before a reasonably conclusive account can be given.

My intentions regarding the material for the book have also changed somewhat. Had I kept going at the level of the two draft chapters on elastic and inelastic scattering I sent you earlier, it would have become quite massive and heavy on the mathematical side. While retaining the scope of the book, covering all "diffractive processes" in nuclear physics, I propose to give a more "physical" presentation, describing only the main results of the analytical formalism without any details of derivation, and emphasizing the physical interpretation of the formulae. This will not only keep the book within reasonable length but should also much improve the readability of the text for non-specialist readers, in accordance with your advice in your earlier letter. Therefore I trust you will approve of this view.

It will be clear from this account of the development of his plans for the book that he regarded the first five chapters as a necessary introduction to the core of the book. Of these, the first three were found in MSS draft form together with a rather less complete Chapter 4. The most important part of the book is that concerned with the closed formalism for strong absorption processes. He completed the first two chapters, on elastic and inelastic scattering, and these form Chapters 6 and 7. The application of his diffraction formalism to nucleon transfer reactions was published in the *Physical Review*, and this forms Chapter 8. Some of the other applications are described in his many papers listed in the appendix.

In preparing this book for the press I have been fortunate to have the assistance of Professor Mahir Hussein, who collaborated with Professor Frahn for many years, particularly on the analysis of heavy ion reactions. He has read through the text, checked the formulae and made many suggestions concerning the selection of materials. His generous help is greatly appreciated. One of his former students, Miss G. M. Field, has helped with the figures and the text.

This is not the book that Willy Frahn would have written had he lived to bring his plans to fruition. It is, however, our hope that it is a useful account of his unique contribution to modern physics and a worthy memorial to his life's work.

Oxford P. H.
1983

CONTENTS

PART I NUCLEAR DIFFRACTION MODELS

PART I

NUCLEAR DIFFRACTION MODELS

1

DIFFRACTION IN NUCLEAR PHYSICS

1.1 Introduction

Diffraction is one of the fundamental phenomena in quantum physics attesting to the wave nature of material particles. Since the wave properties of particles are those of probability waves, diffraction in quantum theory is basically different from diffraction of classical wave fields familiar from classical optics or acoustics: it is purely a quantum phenomenon which disappears in the classical limit. On the other hand, the range of diffraction phenomena in quantum physics is very much wider than that in classical physics: it extends from coordinate space to the space of any quantal observable. In fact, a quantal diffraction pattern can be defined as a particular probability distribution over the spectrum of a given observable which arises as a response to a certain type of quantal measurement of the canonically conjugate observable. Quantal diffraction phenomena are therefore directly associated with the noncommutability of quantum observables and the reciprocal restrictions imposed on their measurement by the Heisenberg uncertainty relations. To emphasize their physical distinctness from their classical counterpart, and their extended scope implied by the general definition just given, the term 'quantum diffraction' has been introduced (Frahn 1977 [1.1]) as a unifying concept comprising the entire range of diffraction phenomena in quantum physics.

The present work deals with quantum diffraction in nuclear processes. Though first discovered in the scattering of electrons by crystals shortly before the advent of quantum mechanics, and later also observed in the scattering of atoms and molecules by crystal surfaces, quantum diffraction is particularly conspicuous in nuclear collisions. This is because atomic nuclei are extended objects of well-defined size and shape which interact strongly with one another so that energetic nuclear collisions lead to a great variety of reaction products. Nuclear sizes happen to be large compared with the de Broglie wavelength of the relative motion of two colliding nuclei at all energies that are high enough for many reaction channels to be open. Thus the zone of interaction effectively removes a finite portion, comprising many partial waves, of the wave-front representing the relative motion of the colliding nuclei.

In actual fact, of course, this line of reasoning is reversed: from the observed diffraction patterns we conclude the size and shape of nuclei. Indeed, measurement of such patterns as scattering angular distributions has been a most important means of obtaining information about nuclear sizes and shapes. The wave-mechanical description of nuclear collisions is *essential*; the classical particle ('billiard ball') picture is entirely inadequate. The diffraction 'models' take care of essentially 'model-independent' aspects of nuclear scattering and reaction processes.

1.2 The wave-mechanical description of collisions

When two nuclei collide they are either scattered elastically or undergo a reaction in which the internal state of one or both collision partners is changed. In wave mechanics, such processes are described as follows. The incoming particles are represented by plane waves; the interaction between them gives rise to secondary waves whose superposition upon the ingoing wave produces a complicated interference pattern; at large distances from the scattering centre the total wave field consists of the original plane wave plus a set of outgoing spherical waves which represent the reaction products in various final states. The amplitude of an outgoing spherical wave depends on the direction of the emerging particles and on the initial and final wave numbers; its squared modulus, the differential cross-section, measures the probability distribution per unit solid angle of the scattered particles.

In contrast with classical mechanics which describes a collision in terms of the deflection of particles moving along well-defined trajectories, the wave-mechanical picture is analogous to that of diffraction of light waves by material objects in classical optics. Max Born, who was the first to give a quantal formulation of collision processes on the basis of Schrödinger's wave mechanics, described the scattering of an electron by an atom as 'nothing but a "diffraction problem" in which the incoming plane wave is diffracted or scattered by the atom; in place of the boundary conditions used in optics for the description of screens, one has here the potential energy of interaction between atom and electron' (Born 1926 [1.2]). At the same time, in order to establish the connection with the corpuscular description, Born took a fundamental step by proposing his *statistical interpretation of quantal processes*: the intensity of the Schrödinger wave field is to be regarded as the probability distribution representing an ensemble of colliding particles.

Were we to adopt the analogy between light scattering in optics and particle collisions in wave mechanics in its widest sense, the present work would have to cover the entire range of nuclear scattering processes.

However, 'nuclear diffraction scattering' customarily refers to a more limited class of phenomena in nuclear collisions. In order to define its scope let us briefly recall the two meanings in which the concept of diffraction is used in classical optics.

The general concept is that of 'rigorous diffraction theory' (see, for instance, Born and Wolf (1970) [1.3]), which formulates diffraction as a *boundary value problem* of the Helmholtz equation (for acoustic waves) or of Maxwell's equations (for electromagnetic waves). This formulation is mathematically well defined and yields, in principle, a unique solution for the total wave field everywhere in space and for any wave number k and linear dimension $\mathbf{a}$ of the diffracting object. It encompasses all scattering situations between the static (Rayleigh) limit $ka \to 0$ and the limit of geometrical optics $ka \to \infty$.

Prior to its formulation as a boundary value problem, however, diffraction was understood in the sense of the *Fresnel–Kirchhoff theory*, which is valid for wavelengths that are small compared with the size of the scatterer, that is for $ka \gg 1$. In this theory, diffraction is described as the modification of the wave field resulting from the *limitation of the wave-front* by the diffracting object; it is an approximation valid at field points not too close to the object and not too far from the geometrical shadow boundary.

In what follows we shall speak of diffraction only in this restricted sense, adopting as the general *diffraction condition* the short-wavelength condition

$$ka \gg 1, \qquad (1.2.1)$$

and refer to the general case as 'scattering'. Thus the relation between nuclear diffraction and nuclear scattering in general will be understood as analogous to the relation between the Fresnel–Kirchhoff approximation and rigorous diffraction theory in classical optics.

The limited concept of diffraction has important advantages to make up for its approximate nature. These are already evident in the optical case. Firstly, exact solutions of the classical boundary value problem are possible only in a very few special cases, for objects of high symmetry (half-plane, sphere, circular disc) and idealized physical properties (perfect conductivity, sharp boundaries, infinitely thin screens). Even if the solution is obtainable in closed form, i.e. in terms of special functions pertaining to the symmetry of the problem, the explicit evaluation involves a series expansion in powers of ka which converges slowly in the short-wavelength region. Evaluation for $ka \gg 1$ requires transformation of the series into an asymptotic expansion in inverse powers of ka which has been achieved only in a few instances. In all other cases one has to resort to numerical computation or to rely on approximation methods such as Kirchhoff's theory. Similarly, nuclear scattering problems can only be solved numerically, except for the simplest forms of interaction potential.

Secondly, the approximate theory can easily deal with more general diffraction problems. The scattering objects may have shapes of lesser symmetry, or diffuse boundaries, or may be composed of several components in regular or irregular spatial arrangement. A scatterer may have both absorptive and refractive properties, affecting both the amplitude ('amplitude object') and the phase ('phase object') of the incoming wave. The wave field may have arbitrary coherence and polarization properties and any given frequency spectrum, and so on. The simple mathematical form of the Fresnel–Kirchhoff diffraction theory, in which the scattering amplitude is essentially the (two-dimensional) Fourier transform of the object function, allows one to establish simple relations between the properties of the scatterer and those of the diffraction pattern it produces. Many of the developments in modern physical optics ('Fourier optics', see e.g. Goodman (1968) [1.4]) are thus based on the restricted, Fresnel–Kirchhoff concept of diffraction.

1.3 Quantum diffraction

After having highlighted the close formal similarity between diffraction in wave-mechanical scattering theory and diffraction in classical optics, we must emphasize a fundamental physical difference. According to Born's statistical interpretation, the wave fields involved in the quantal description of a collision process are probability waves whose interference and diffraction is governed by the superposition principle of quantum mechanics. Quantal interference and diffraction patterns are therefore determined by the quantum of action $\hbar$ and *vanish in the classical limit* $\hbar \to 0$, while the phenomena of interference and diffraction in classical optics are independent of $\hbar$. It appears, therefore, that diffraction in wave-mechanical processes is basically associated with the quantal limitations of physical measurements expressed by Heisenberg's uncertainty principle, i.e. with the existence of incompatible observables.

Thus we are led to a generalized concept of diffraction in quantum theory, which we call '*quantum diffraction*' (Frahn 1977 [1.1]). In this view the quantal analogue of diffraction in classical optics is a consequence of the uncertainty relation between linear coordinates and their canonically conjugate linear momenta. In addition, however, there are diffraction phenomena associated with any other pair of canonically conjugate observables, such as energy and time, angular momentum and angle, or particle number and phase. Since collision processes are described in terms of scattering amplitudes and differential cross-sections as functions of scattering angle, which are appropriately represented as expansions in eigenfunctions of angular momentum ('partial-wave expansion'), the quantal diffraction phenomena associated with orbital angular momentum and scattering angle will be of particular importance for our study of diffractive effects in nuclear collisions.

1.4 Diffraction in nuclear processes

The evidence for diffraction phenomena in nuclear interactions comes primarily from experiments measuring the angular distributions for elastic nuclear scattering and certain nuclear reactions. Many of these distributions show a regular pattern of oscillations whose period and amplitude change with wave number in much the same way as in a Fraunhofer diffraction pattern in classical optics. Examples are shown in Fig. 1.1. Except in neutron scattering, all nuclear collisons involve the Coulomb repulsion between the nuclear charges which modifies the angular distributions resulting from the nuclear interaction in a characteristic fashion. We shall see that with increasing strength of the Coulomb interaction the character of nuclear diffraction scattering changes from Fraunhofer type to Fresnel type. The latter is particularly conspicuous in the scattering of heavy ions, an example of which is shown in Fig. 1.2.

Why are diffractive effects so prominent in nuclear collisions, in contrast with, for instance, atomic or molecular scattering? There are two main reasons. Firstly, nuclei have a spatial extension of quite well-defined size and shape. Secondly, nuclear sizes happen to be large compared with the de Broglie wavelength of the relative motion of two colliding nuclei at all energies high enough to produce a variety of nuclear reactions. Both these features are due to the characteristics of the nuclear force: its great strength, short range, and saturation properties. Thus whenever the diffraction condition (1.1) is satisfied there is usually also strong *absorption*, defined as the depletion of particle flux in the elastic channel owing to presence of open nonelastic channels. Such conditions are analogous to those for diffraction of light by opaque or only partially transparent objects with well-defined boundaries in classical optics.

Nuclear diffraction is by no means confined to elastic scattering. Many nonelastic processes, in particular inelastic scattering and direct reactions such as quasi-elastic scattering and nucleon transfer reactions, are dominated by diffractive effects. In fact, such effects are present whenever a sufficient number of partial waves pertaining to a 'band' of orbital angular momenta participates coherently in the interaction.

Simple basic diffraction patterns are, however, usually modified to a

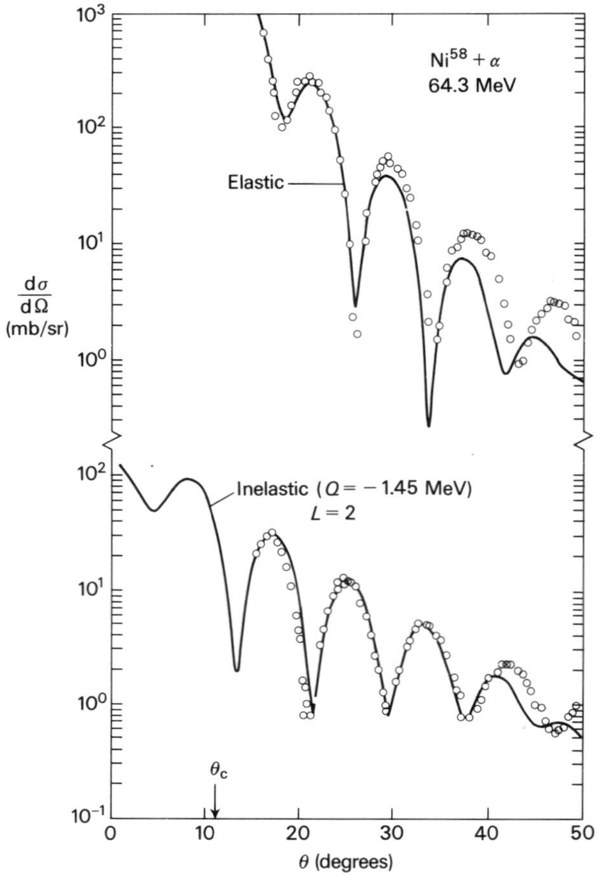

FIG. 1.1. Differential cross-section for elastic and inelastic scattering of 64.3 MeV alpha particles by ^{58}Ni. The curve for the elastic angular distribution is calculated by means of the SAM formula (with $\varepsilon_1 = \varepsilon_2 = 0$), and the curve for the inelastic angular distribution by means of eqn (7.3.12) with the form factor (6.3.17). The parameters are $T = 22.67$, $\Delta = 1.17$, $\mu = 1.50$ *and* $\delta_2 = 0.94$ fm. The Coulomb parameter is $n = 2.20$ and the critical angle $\theta_c = 11.1°$. (From Potgieter and Frahn *Nucl. Phys.* **80,** 434 (1966).)

lesser or greater extent by nondiffractive effects of the interaction. Firstly there are those global properties of nuclear structure which, though well within the scope of diffraction theory, cause deviations from the identical picture of the nucleus as a perfectly absorbing sphere with sharp boundary: in addition to the Coulomb interaction whose strength determines the *type* of diffraction scattering, there are modifications due to diffuseness of the nuclear surface, deformation of the spherical shape, finite

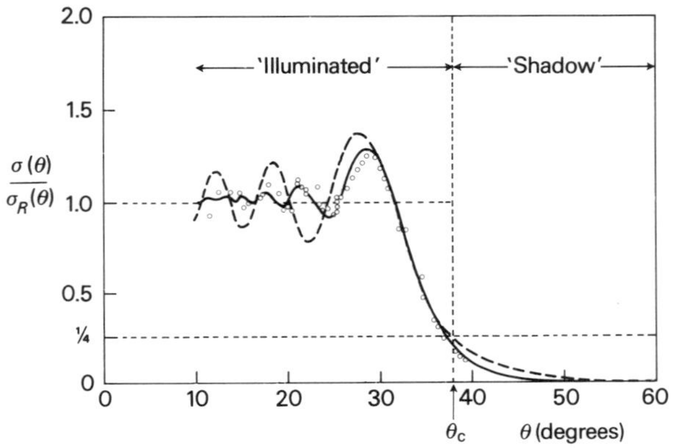

FIG. 1.2. Differential cross-section for the elastic scattering of 170.1 MeV ^{16}O by ^{208}Pb (Baker and McIntyre 1967) showing a typical Fresnel diffraction pattern. The solid curve is obtained from the strong absorption model (Frahn and Venter 1963) and the broken curve is the Fresnel cross-section calculated from eqn (4.7.21) with $n = 31.2$ and $\Lambda = 90.4$. This comparison shows the extent to which the Fresnel cross-section is retained in spite of the effects of surface diffuseness ($\Delta = 3.41$) and real nuclear phase shifts ($\mu/4\Delta = 0.33$). (Frahn, W. E. (1971). *Phys. Rev. Lett.* **26,** 568; (1972). *Ann. Phys. (NY)* **72,** 524.)

transparency and refractive properties of the nuclear interior, as well as the effects of spin (polarization), isospin, and the symmetry properties of the wave function in the case of identical nuclei. Among the 'refractive' properties of the interaction are included the typical effects of potential scattering, such as rainbow and glory scattering. Secondly there are the important modifications due to specific structure properties of individual nuclei: the effects of shell structure, pairing and superfluidity, etc. The presence of resonances and other effects associated with individual partial waves rather than with a coherent band of angular momenta usually affects the angular distribution mainly at backward angles, and shows up most specifically in the energy dependence of the differential cross-section (excitation function).

1.5 References

[1.1] Frahn, W. E. (1979). *Quantum diffraction. Revista del Nuovo Cimento* **7,** 499.

[1.2] Born, M. (1926). *Z. Physik* **37,** 863.

[1.3] Born, M. and Wolf, E. (1970). *Principles of optics* (4th edn), Pergamon, Oxford.

[1.4] Goodman, J. W. (1968). *Introduction to Fourier optics.* McGraw-Hill, New York.

2

THE DEVELOPMENT OF NUCLEAR DIFFRACTION

2.1 Early beginnings

As we have seen, nuclear diffraction is a wave-mechanical phenomenon observed in nuclear collision processes which is associated with the finite size and absorptive properties of nuclei. Let us now retrace its historical development.

In 1919, Rutherford, during experiments to test his nuclear model of the atom, made two discoveries which gave the first indication of finite nuclear size and of nuclear absorption. Firstly he found (Rutherford 1919 [2.1]) that the distribution of alpha particles scattered by hydrogen at large angles deviated from the calculated angular distribution for a Coulomb interaction between point particles. Subsequent experiments on hydrogen and other light nuclei (He, Al, Mg) showed a marked decrease of the scattering cross-section $\sigma(\theta)$ below the Rutherford distribution $\sigma_R(\theta)$ at angles corresponding to distances of approach smaller than 4 to 8 times 10^{-13} cm. This 'anomalous scattering' was attributed by Bieler (1923 [2.2]) to an attractive nuclear force acting in addition to the repulsive Coulomb force at small distances. To explain the observed angular distributions, various models of the alpha–nucleus interaction were proposed which predicted most of the phenomena we are now familiar with in potential scattering, such as rainbow scattering, spiralling, orbiting, etc. (see Ford and Wheeler (1959) [2.44]). For instance, the calculations of Debye and Hardmeier (1926 [2.3]) for an alpha–nucleus potential of the form

$$V(r) = \frac{2Ze^2}{r} - \frac{a}{r^n} \tag{2.1.1}$$

with $n = 4$ showed that particles moving on trajectories with large impact parameters b are scattered through angles that increase with decreasing b up to a maximum angle of deflection (now called the 'rainbow angle'). When b becomes smaller than a critical value b_0 (corresponding to 'orbiting'), the alpha particles may again emerge at all angles on trajectories that wind around the scattering centre; finally, at still smaller b

values the particles spiral towards the centre to be captured by the target nucleus.

The possibility of alpha particle capture ('absorption') resulted from Rutherford's (1919 [2.4]) discovery of the first nuclear reaction, when he observed long-range protons emerging from alpha collisions with nitrogen nuclei. Such processes were also considered to contribute to the deviation of the angular distribution for elastic scattering from $\sigma_R(\theta)$ at large angles. Experiments during the following decade on alpha scattering and alpha-induced reactions with light target nuclei established that the nucleus has a fairly well-defined radius that increases with mass number as $R = r_0 A^{1/3}$, where $r_0 = 1.4\text{–}1.5 \times 10^{-13}$ cm, and a high probability of capture expressed by a reaction cross-section of the order $\sigma_r \approx \pi R^2$ (Pollard 1935 [2.5]).

Meanwhile, quantum mechanics had evolved as a new fundamental theory of atomic processes. It predicted marked deviations from classical motion if the de Broglie wavelength λ becomes comparable with the size of the scatterer. Since the alpha particles in Rutherford's collision experiments have wavelengths of the order of 10^{-13} cm, only a few times smaller than the nuclear radius R, 'one may with reason inquire whether the classical mechanics is adequate to deal with these very close collisions and whether we should not rather adopt the methods of the new wave mechanics' (Rutherford *et al.* (1930) [2.6], p. 276).

The first wave-mechanical formulation of particle collisions was given by Born (1926 [2.7]), who described it as 'nothing but a diffraction problem'. (Elsasser (1925) [2.7a] was the first to conclude that on the basis of de Broglie's theory, the collision of particles must be accompanied by diffraction effects.) Criteria for the conditions under which such diffraction effects may be expected to appear in particle collisions had been given by de Broglie (1926 [2.8]) and Schrödinger (1926 [2.9]). These were applied to alpha–nucleus scattering by Blackett (1927 [2.10]) with the result that the deviation from classical behaviour is essentially determined by the quantity $n = Z_1 Z_2 e^2/\hbar v$ (now called the 'Sommerfeld parameter'): the classical description is valid to a good approximation only if $n \gg 1$. In Rutherford's experiments the values of n range from about 0.2 to 3.7. Yet when Born's wave-mechanical theory was first applied to alpha–nucleus scattering with an inverse square Coulomb potential by Wentzel (1926 [2.11]) and Oppenheimer (1927 [2.11a]) for small n and later in partial-wave expansion for general n by Mott (1928 [2.12]), the quantal angular distribution turned out to be exactly equal to the classical result $\sigma_R(\theta)$! Also the calculations of Banerji (1930 [2.13]), Massey (1930 [2.14]), Sexl (1930 [2.15]), and others for interaction potentials of the form (2.1) showed no essential difference between the quantal and the corresponding classical results. Thus, proper diffraction effects were expected to obtain only from higher-order corrections to the Born approximation, and it was not until much later that alternative wave-mechanical descriptions of collision processes, proposed by Faxén and Holtsmark (1927 [2.16]) for electron scattering from atoms, were applied to nuclear collisions. In view of this unsettled situation, Rutherford *et al.* (1930 [2.6], p. 277) concluded that 'it is difficult to suggest where diffraction patterns may be expected to appear in collisions of α particles with atomic nuclei'.

In their experiments, Rutherford and his coworkers had to use alpha particles from natural radioactive sources with energies less than about 9 MeV. The exploration of larger deviations from Coulomb scattering, to be expected in collisions at closer distances of approach and from heavier target nuclei, had to await the invention of particle accelerators. Although the first cyclotrons were developed around 1930, systematic scattering experiments with cyclotron-produced alpha particle beams were not made until 20 years later.

Meanwhile, a new window on the nucleus was opened by Chadwick's discovery of the neutron in 1932. This was immediately followed by Heisenberg's theory of nuclear constitution with protons and neutrons as basic components. Within a few years nuclear models were proposed, by Elsasser, Guggenheimer, Margenau, Heisenberg, and others, which pictured the nucleus in analogy to atomic structure as a system of quasi-independent nucleons in a common potential well and suggested a shell structure of light nuclei. Parallel with these advances in nuclear theory, intense activity in the study of new types of nuclear reaction was initiated by the discovery by Fermi and his coworkers in 1934 of the large cross-sections for capture of slow neutrons by nuclei.

The scattering of neutrons by nuclei was first described by Bethe (1935 [2.17]) and Fay (1936 [2.18]) who represented the neutron–nucleus interaction by a real single-particle potential well. To account for the presence of neutron-induced reactions, Ostrofsky *et al.* (1936 [2.19]) introduced a complex potential well of which the imaginary part represents nuclear absorption.

On the other hand, the results of Fermi's group on the high probability of slow-neutron capture led Bohr (1936 [2.20]) to emphasize much more strongly the absorptive properties of the interaction, to such an extent that the nucleus is regarded as a perfect absorber. In this picture the capture of a slow neutron results in the formation of a long-lived intermediate system, the 'compound nucleus', whose eventual decay is independent of the way it was formed. If the interaction is described by a complex potential, the strong absorption can be represented by a large imaginary part; in fact Bethe (1940 [2.21]) considered the extreme of a purely imaginary potential.

It was soon realized that perfect absorption is always associated with elastic scattering ('shadow scattering'). This important relation, which later became known as the 'optical theorem', seems to have been first proposed by Bohr *et al.* (1940 [2.22]). If the nucleus is represented by a completely absorbing sphere of radius R, the differential cross-section for elastic scattering has a form, given by Placzek and Bethe (1939 [2.23]), that is analogous to the Airy pattern of diffraction from a circular aperture:

$$\sigma(\theta)=\left\{\frac{RJ_1(kR\theta)}{\theta}\right\}^2, \tag{2.1.2}$$

where $J_1(x)$ denotes a cylindrical Bessel function of first order. Thus the angular distribution for elastic neutron scattering has the same form as the Fraunhofer diffraction pattern of light scattered by a completely opaque ('black') disc of radius R. This result is the first prediction of pure diffraction scattering in nuclear physics.

Formula (2.1.2) appeared to be in qualitative agreement with experiments that indicated the existence of elastic neutron scattering at small angles. Although subsequent experiments confirmed the presence of an oscillatory structure in angular distributions at higher energies, these patterns were not in agreement with the assumption of complete absorption. Following a suggestion of Serber (1947 [2.24]) that nuclei are partially transparent to high-energy neutrons, Fernbach *et al.* (1949 [2.25]) re-introduced a real part in the neutron–nucleus interaction and analysed the data of Cook *et al.* (1949 [2.26]) for 90 MeV neutrons by a complex square-well potential corresponding to a sphere of complex refractive index in optics. (This is the origin of the term 'optical model'). A similar description was given by Pasternak and Snyder (1950 [2.27]).

By 1945, Akhiezer and Pomeranchuk (1945 [2.28]) had already considered the assumption of perfect absorption in the scattering of *charged* particles and had derived a modification of the Fraunhofer diffraction pattern (2.1.2) due to a (weak) Coulomb interaction:

$$\sigma(\theta)=\left[\frac{RJ_1\{kR(1-2n/kR)^{1/2}\theta\}}{\theta}\right]^2 \tag{2.1.3}$$

where n is the Sommerfeld parameter. The complex potential model was applied to charged-particle scattering at lower energies, first by Le Levier and Saxon (1952 [2.29]) for 18 MeV protons and later by Chase and Rohrlich (1954 [2.30]), Woods and Saxon (1954 [2.31]), and many others in the following years. In all cases it was found that the angular distributions had more or less regular oscillatory structure indicative of diffraction scattering, but deviated from the ideal Fraunhofer pattern by appreciable nuclear transparency, diffuseness of the nuclear surface, Coulomb and spin effects, etc. By ever-increasing generalizations of the complex nuclear interaction potential the 'optical model' has been very successfully adapted to describe all types of nuclear scattering processes up to the present time.

2.2 Analyses of alpha particle scattering

In the early 1950s the study of alpha particle scattering was resumed with cyclotron-produced alpha beams. The earliest of these experiments were carried out by Gove in 1950 with 27.5 MeV alphas from the Massachusetts Institute of Technology cyclotron, but the results were not published until 1955 (Gove (1955) [2.34]). The first systematic studies were made by Farwell and Wegner (1954 [2.35]) at the University of Washington cyclotron with alpha particles in the range 13–40 MeV scattered from heavy nuclei. These were followed by the data of Wall *et al.* (1955 [2.36]) for 22 MeV alpha scattering by Ag, Au, and Pb. The observed angular distribution showed strong deviations from Rutherford scattering at all but small angles: after an initial rise above the Rutherford values the cross-section dropped sharply and smoothly below $\sigma_R(\theta)$ with increasing angle.

The first theoretical interpretation of these results was given by Blair (1954 [2.37]). Employing the partial-wave expansion he assumed, similar to Akhiezer and Pomeranchuk (1945 [2.28]), that all partial waves below a critical angular momentum l_c are completely absorbed while all higher partial waves undergo pure Rutherford scattering. The resulting differential cross-section, obtained by numerical summation of the partial-wave series, exhibited strong diffraction-like oscillations which were not seen in the observed angular distributions. However, after averaging over the oscillations the mean cross-section closely approximated the data. Blair noticed that at a critical angle θ_c, the 'grazing angle', pertaining to l_c by the Rutherford scattering relation $l_c+\frac{1}{2}=n\cot\frac{1}{2}\theta_c$, the cross-section is always close to a quarter of the Rutherford cross-section. This property can be used to determine the only parameter in the theory, l_c, and hence the nuclear radius R by virtue of the semiclassical relation between the scattering angle and the distance of closest approach ('one-quarter-point recipe').

Blair himself had already surmised that the appearance of the strong Fraunhofer-type oscillations is due to the assumption of an abrupt change from complete absorption to full transmission at l_c ('sharp cut-off assumption'), and will be smoothed out by a gradual transition in l-space. After an earlier attempt by Cheston and Glassgold (1957 [2.38]) to modify the Blair model in this respect, the first successful 'smooth cut-off' model was given by McIntyre *et al.* (1960 [2.39]) who assumed a particular functional form for the gradual change of η_l from complete absorption a low l values to full transmission of large-l partial waves. In analysing the alpha scattering data of Wegner *et al.* (1955 [2.40]) at higher energy (40 MeV), which did show the presence of regular diffraction-like oscillations at larger angles (although strongly damped compared with the Blair model prediction), McIntyre *et al.* found it necessary to introduce another extension of the Blair model: satisfactory agreement with the data could only be obtained by assuming that the partial waves undergo nuclear phase shifts in addition to the phase shifts caused by Rutherford scattering.

At about the same time, Greider and Glassgold (1960 [2.41]) applied various functional parametrizations of η_l to the scattering of high-energy neutrons (e.g. De Juren 1950; Atkinson *et al.* 1959 [2.42]). They succeeded in obtaining closed expressions of the scattering amplitude by approximate analytic summation of the partial-wave series.

Alternative descriptions of alpha–nucleus scattering had been proposed based on the consideration that, for heavy particles moving in strong Coulomb fields, classical mechanics may be a fair approximation to quantal motion. In view of the observation by Wegner that the angle and energy dependence of alpha scattering cross-sections can be represented in an almost universal way as a function of distance of closest approach ('apsidal distance'), Porter (1955 [2.43]) proposed a 'semiclassical' model in which the deviation from Rutherford scattering is described by an attenuation factor which is calculated as the probability of absorption of the alpha particles along classical trajectories. This factor is shown to depend essentially on the ratio D/R of apsidal distance to nuclear radius. In a seminal paper, Ford and Wheeler (1959 [2.44]) made a thorough investigation of the phenomena that can appear in a semiclassical description of potential scattering (rainbow scattering, glory scattering, orbiting, and spiralling) and elucidated the close relations with the corresponding phenomena in light optics. The basic quantity in the Ford–Wheeler theory is the 'deflection function', i.e. the derivative of the scattering phase shift with respect to orbital angular momentum, which is calculated from a given interaction potential by a semiclassical (Jefferies–Wentzel–Kramers–Brillouin (JWKB)) approximation. Ford and Wheeler succeeded in deriving closed expressions for the cross-sections of the various types of scattering by evaluating the partial-wave summations in stationary phase approximation. Despite the physical importance of the work of Ford and Wheeler it is fair to say that its application to alpha–nucleus scattering was not entirely successful. Since absorption is neglected or assumed to be weak, the model fails to reproduce the characteristic diffraction effects in nuclear scattering. The assumptions underlying Porter's model also correspond to relatively weak absorption. As a result, the finer details associated with quantal diffraction which are observed in most alpha scattering cross-sections are not reproduced. Both types of semiclassical description have been revived quite recently in heavy-ion scattering.

2.3 References

[2.1] Rutherford, E. (1919). *Phil. Mag.* **37,** 537.
[2.2] Bieler, E. S. (1923). *Proc. Camb. Phil. Soc.* **21,** 686.
[2.3] Debye, P. and Hardmeier, W. (1926). *Physik. Z.* **27,** 196; Hardmeier, W. (1926). *ibid* **27,** 574.
[2.4] Rutherford, E. (1919). *Phil. Mag.* **37,** 581.
[2.5] Pollard, E. C. (1935). *Phys. Rev.* **47,** 611.
[2.6] Rutherford, E., Chadwick, J., and Ellis, C. D. (1930). *Radiations from radioactive substances.* Cambridge University Press, Cambridge.
[2.7] Born, M. (1926). *Z. Physik* **37,** 863; *ibid* **38** (1926), 803; *Göttingen Nachrichten* (1926), p. 146.
[2.7a] Elsasser, W. (1925). *Naturwiss.* **13,** 711.
[2.8] de Broglie, L. (1926). *J. de Physique et le Radium* **7,** 321.
[2.9] Schrödinger, E. (1926). *Z. Physik* **79,** 489.
[2.10] Blackett, P. M. S. (1927). *Proc. Camb. Phil. Soc.* **23,** 698.
[2.11] Wentzel, G. (1926). *Z. Physik* **40,** 590.
[2.11a] Oppenheimer, J. R. (1927). *Z. Physik* **43,** 413.
[2.12] Mott, N. F. (1928). *Proc. R. Soc. (Lond.)* **A118,** 542.
[2.13] Banerji, A. C. (1930). *Phil. Mag.* **9,** 273.
[2.14] Massey, H. S. W. (1930). *Proc. R. Soc. (Lond.)* **A127,** 671.
[2.15] Sexl, T. (1930). *Naturwiss.* **18,** 247.
[2.16] Faxén, H. and Holtsmark, J. (1927). *Z. Physik* **45,** 307.
[2.17] Bethe, H. A. (1935). *Phys. Rev.* **47,** 747.
[2.18] Fay, C. H. (1936). *Phys. Rev.* **50,** 560.
[2.19] Ostrofsky, M., Breit, G., and Johnson, D. P. (1936) *Phys. Rev.* **49,** 22.
[2.20] Bohr, N. (1936). *Nature (Lond.)* **137,** 344, 351.
[2.21] Bethe, H. A. (1940). *Phys. Rev.* **57,** 1125.
[2.22] Bohr, N., Peierls, R. E., and Placzek, G. (1939). *Nature (Lond.)* **144,** 200.
[2.23] Placzek, G. and Bethe, H. A. (1940). *Phys. Rev.* **57,** 1075.
[2.24] Serber, R. (1947). *Phys. Rev.* **72,** 1114.
[2.25] Fernbach, S., Serber, R., and Taylor, T. B. (1949). *Phys. Rev.* **75,** 1352.
[2.26] Cook, L. J., McMillan, E. M., Peterson, J. M., and Sewell, D. C. (1949). *Phys. Rev.* **75,** 7.
[2.27] Pasternak, S. and Snyder, H. S. (1950). *Phys. Rev.* **80,** 921.

[2.28] Akhiezer, A. and Pomeranchuk, I. (1945). *J. Phys.* (*Moscow*) **9,** 471.
[2.29] Le Levier, R. E. and Saxon, D. S. (1952). *Phys. Rev.* **87,** 40.
[2.30] Chase, D. M. and Rohrlich, F. (1954). *Phys. Rev.* **94,** 81.
[2.31] Woods, R. W. and Saxon, D. S. (1954). *Phys. Rev.* **95,** 577.
[2.32] Melkanoff, M. A., Moszkowski, S. A., Nodvik, J., and Saxon, D. S. (1956). *Phys. Rev.* **101,** 507.
[2.33] Glassgold, A. E., Cheston, W. B., Lewis, M. L., Schneldt, S. B., and Erickson, G. W. (1957). *Phys. Rev.* **106,** 1207.
[2.34] Gove, H. E. (1955). *Phys. Rev.* **99,** 1353.
[2.35] Farwell, G. and Wegner, H. E. (1954). *Phys. Rev.* **95,** 1212.
[2.36] Wall, N. S., Rees, J. R., and Ford, K. W. (1955). *Phys. Rev.* **97,** 726.
[2.37] Blair, J. S. (1954). *Phys. Rev.* **95,** 1218.
[2.38] Cheston, W. B. and Glassgold, A. E. (1957). *Phys. Rev.* **106,** 1215.
[2.39] McIntyre, J. A., Wang, K. H., and Becker, L. C. (1960). *Phys. Rev.* **117,** 1337.
[2.40] Wegner, H. E., Eisberg, R. M., and Igo, G. (1955). *Phys. Rev.* **99,** 825.
[2.41] Greider, K. R. and Glassgold, A. E. (1960). *Ann. Phys.* (*N.Y.*) **10,** 100.
[2.42] De Juren, J. (1950). *Phys. Rev.* **80,** 27. Atkinson, J. H., Hess, W. N., Perez-Mendez, V., and Wallace, R. W. (1959). *Phys. Rev. Lett.* **2,** 168.
[2.43] Porter, C. E. (1955). *Phys. Rev.* **99,** 1400.
[2.44] Ford, K. W. and Wheeler, J. A. (1959). *Ann. Phys.* (*N.Y.*) **7,** 259.

3

THE THEORY OF DIFFRACTION SCATTERING

3.1 The quantal description of collisions

In this chapter we discuss the quantal description of collisions (in so far as it is needed for our discussion of diffraction scattering), giving particular attention to the aspect of diffraction.

In a scattering experiment a beam of particles produced by a source, usually an accelerator with a collimating system, impinges on a target; particles with specified properties emerging from the target in a given direction are counted by a detector. Both the beam and the target are ensembles of individual particles. Nuclear scattering experiments are normally conducted under conditions such that the members of these ensembles undergo independent interactions. The probability of a collision can then be expressed by a *cross-section*; this represents the primary observable quantity from which all information about the collision process is derived by means of the quantum theory of scattering.

The scattering situation can be represented by a wave function $\psi^{(+)}(\mathbf{k}, \mathbf{r})$ describing the relative motion of the collision partners. Since the scattered particles are detected at a distance that is large compared with the size of the interaction region, the cross-section depends only on the asymptotic form of the wave function in coordinate space; in the simplest case of elastic scattering of spinless particles this consists of a superposition of an incoming plane wave and a radially outgoing (spherical) scattered wave:

$$\psi^{(+)}(\mathbf{k}, \mathbf{r}) \simeq \mathrm{e}^{\mathrm{i}\mathbf{k}\cdot\mathbf{r}} + f(\mathbf{k}, \hat{\mathbf{r}}) \frac{\mathrm{e}^{ikr}}{r}. \tag{3.1.1}$$

The amplitude of the spherical wave, which depends on the wave vector $\mathbf{k}$ of the incoming wave and on the direction of observation $\hat{\mathbf{r}}$, determines the differential scattering cross-section by

$$\frac{\mathrm{d}\sigma}{\mathrm{d}\Omega} = |f(\mathbf{k}, \hat{\mathbf{r}})|^2. \tag{3.1.2}$$

The wave function $\psi^{(+)}(\mathbf{k}, \mathbf{r})$ describes a time-independent stationary-state situation quite unlike the actual collision process involving the motion of the collision partners within a finite time interval between production of the projectile and detection of the outgoing particle. An appropriate wave-mechanical description would be in terms of wave packets of finite extension and duration that evolve according to the dynamics of the time-dependent Schrödinger equation. The question as to why the stationary-state scattering formalism is nevertheless an excellent approximation for most nuclear and atomic collision problems has been carefully discussed in the literature (see especially Goldberger and Watson (1964) [3.1], Messiah (1961) [3.2], Gottfried (1966) [3.3], and Austern (1970) [3.4]), so there is no need to reproduce the argument in detail here. However, since we are concerned with quantum diffraction phenomena in collision processes, which are associated with spatial and temporal limitations of wave forms, it may be in order to briefly summarize the conditions under which the stationary-state description is valid. The point is, in short, that the diffractive effects resulting from the finite extension and duration of a wave packet representing an accelerator-produced particle are negligibly small because the characteristic lengths involved (diameter of the collimating aperture, width of the wave packet, and the distances between the source and the detector) are very large (macroscopically) compared with the dimensions of the interaction region.

The situation is illustrated in Fig. 3.1(a). The longitudinal and transverse (with respect to $\mathbf{k}$) dimensions of the wave packet are determined by the conditions of its formation in the accelerator: the (longitudinal) width of the packet is $W \approx (\hbar k/m)\tau$, where τ is the characteristic time of production of a particle ($\tau \approx 10^{-10}$ s), and its (transverse) breadth B is defined by the radius of the collimating aperture (typically $B \approx 10^{-3}$ m). Such wave packets do not change their form appreciably over distances of the order of the reduced wavelength λ. They are 'wide and broad':

$$W \gg \lambda; \qquad B \gg \lambda. \tag{3.1.3}$$

From the uncertainty principle it then follows that the spread of longitudinal and transverse momenta is small:

$$\Delta k_{\parallel} \approx W^{-1} \ll k; \qquad \Delta K_{\perp} \approx B^{-1} \ll k. \tag{3.1.4}$$

In order that the scattering will not depend essentially on the form of the wave packet its dimensions must be large compared with the size of the scatterer:

$$W \gg a, \qquad B \gg a. \tag{3.1.5}$$

The particle beam is an ensemble of wave packets that have different impact parameters b and times of formation. Only those packets for

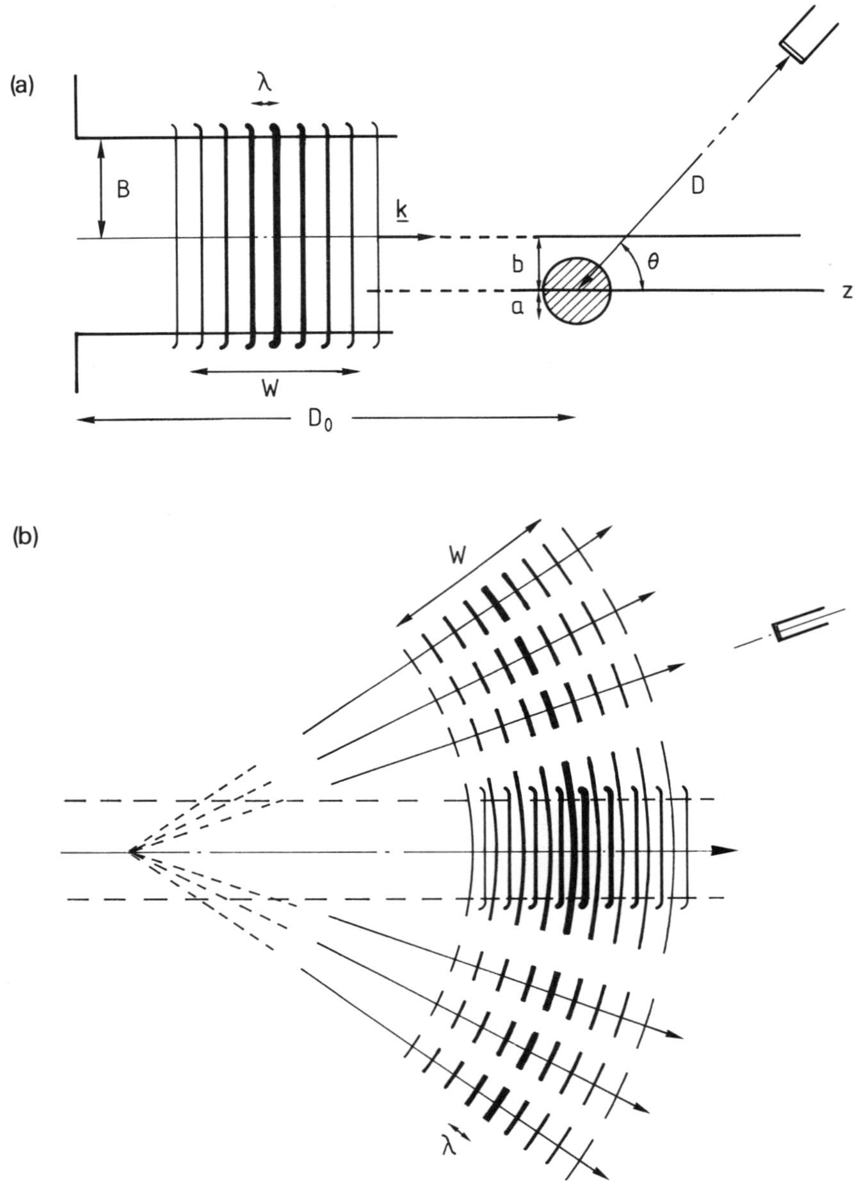

FIG. 3.1. Schematic representation of the scattering process, illustrating the effect of collision on the nature of the incident wavepacket (a) before and (b) after collision.

which $b<B$ can undergo scattering. After the collision (Fig. 3.1(b)) the total wave packet at large distances is a superposition of a spherical packet representing the scattered particle and the ongoing unscattered wave. For the detector to be able to distinguish between these components we must require that $B \ll D \sin\theta$, where D is the distance of the detector and θ is the scattering angle; though there is always a range of angles around the forward direction for which this condition cannot be satisfied, we must ensure that $D \gg B$. Furthermore, in order that the incoming wave packet is initially noninteracting it must be formed at a distance D_0 from the scatterer that is much longer than its width W; thus we require

$$W \ll D_0, \qquad B \ll D. \tag{3.1.6}$$

On the other hand, the distances D_0 and D must not be so large that there is appreciable dispersive spreading of the wave packets over these distances owing to the finite spread in the momentum distribution. For spreading to be negligible we must require

$$D_0 \quad \text{and} \quad D \ll \frac{k}{(\Delta k_{\|})^2} \approx \frac{2W^2}{\lambda} \tag{3.1.7}$$

In summary we have

transverse conditions:

$$\lambda \quad \text{and } a \ll B \ll D \ll 2W^2/\lambda, \qquad b<B,$$

longitudinal conditions: (3.1.8)

$$\lambda \quad \text{and} \quad a \ll W \ll D_0 \ll 2W^2/\lambda.$$

For nuclear collisions, $a \approx 10^{-14}$ m and λ is smaller than 5×10^{-15} m (the value for a 1 MeV proton); B is typically 10^{-3} m and W is of the same order or larger, while D and D_0 are of the order of 1 m. Thus, $2W^2/\lambda \gtrsim 4\times10^{9}$ m (!).

Since our conditions are so amply fulfilled, the motion of the wave packets can be described by the displacements of their centres according to classical mechanics with no detectable change in their form. The differential cross-section is then calculated from the flux in the scattered wave packet with amplitude $f(\mathbf{k}, \hat{\mathbf{r}})$ divided by the flux density in the incoming wave packet with the same result (3.1.2) as for stationary-state wave functions of the form (3.1.1). All we shall be concerned with is the relative magnitude of λ, a, and b which is not restricted by wave packet considerations as long as conditions (3.1.8) are satisfied.

3.2. The S-matrix

Consider a collision process with two particles in the initial state and two particles in the final state

$$\begin{array}{lccc} & a_1+a_2 & \rightarrow & b_1+b_2 \\ \text{(mode of fragmentation)} & a & & b \\ \text{(channel)} & \alpha & & \beta \end{array} \tag{3.2.1}$$

We describe the scattering in the centre-of-mass (c.m.) system.

The *initial channel* α is characterized by

break-up a, momenta $\mathbf{k}_a$ and $-\mathbf{k}_a$

spins $(\mathbf{s}_1, m_1)$ and $(\mathbf{s}_2, m_2)$ couple to channel spin

$$\mathbf{s}=\mathbf{s}_1+\mathbf{s}_2, \qquad m_s=m_1+m_2.$$

The *final channel* β is characterized by

break-up b, momenta $\mathbf{k}_b$ and $-\mathbf{k}_b$

spins $(\mathbf{s}_1', m_1')$ and $(\mathbf{s}_2', m_2')$ couple to channel spin

$$\mathbf{s}'=\mathbf{s}_1'+\mathbf{s}_2', \qquad m_s'=m_1'+m_2'.$$

The asymptotic form of the wave function is

$$\boldsymbol{\psi}_\alpha^{(+)} \approx \exp(\mathrm{i}\mathbf{k}_a \cdot \mathbf{r}_a)\phi_\alpha + \sum_\beta f_{\beta\alpha}(\hat{\mathbf{k}}_b, \hat{\mathbf{k}}_a)\phi_\beta \frac{\exp(\mathrm{i}k_b r_b)}{r_b} \tag{3.2.2}$$

where $\hat{\mathbf{k}}_b=\hat{\mathbf{r}}_b$ and

$$\phi_\alpha=\{\phi_{\alpha_1}\phi_{\alpha_2}\}_\alpha, \qquad \phi_\beta=\{\phi_{\beta_1}\phi_{\beta_2}\}_\beta. \tag{3.2.3}$$

The differential cross-section for the reaction $\alpha\rightarrow\beta$ is

$$\frac{\mathrm{d}\sigma_{\beta\alpha}}{\mathrm{d}\Delta_b}(\hat{\mathbf{k}}_b, \hat{\mathbf{k}}_a)=\frac{v_b}{v_a}|f_{\beta\alpha}(\hat{\mathbf{k}}_b, \hat{\mathbf{k}}_a)|^2 \tag{3.2.4}$$

where

$$v_a=\hbar\mathbf{k}_a/\mu_a, \qquad v_b=\hbar\mathbf{k}_b/\mu_b \tag{3.2.5}$$

$$\mu_a=\frac{M_{a_1}M_{a_2}}{M_{a_1}+M_{a_2}}, \qquad \mu_b=\frac{M_{b_1}M_{b_2}}{M_{b_1}+M_{n_2}}. \tag{3.2.6}$$

In terms of the transition matrix (on the energy and momentum shells)

$$T_{\beta\alpha}(\hat{\mathbf{k}}_b, \hat{\mathbf{k}}_a)=-\frac{2\pi\hbar^2}{\mu_b}f_{\beta\alpha}(\hat{\mathbf{k}}_b, \hat{\mathbf{k}}_a). \tag{3.2.7}$$

This becomes

$$\frac{d\sigma_{\beta\alpha}}{d\Omega_b}(\hat{\mathbf{k}}_b, \hat{\mathbf{k}}_a) = \frac{\mu_b\mu_a}{(2\pi\hbar^2)^2}\frac{k_b}{k_a}|T_{\beta\alpha}(\hat{\mathbf{k}}_b, \hat{\mathbf{k}}_a)|^2 \quad (3.2.8)$$

Define the S-matrix (on the energy shell)

$$S_{\beta\alpha}(\hat{\mathbf{k}}_b, \hat{\mathbf{k}}_a) = \delta_{\beta\alpha}\delta(\hat{\mathbf{k}}_b, \hat{\mathbf{k}}_a) - \mathrm{i}\frac{(\mu_b k_b)^{1/2}}{2\pi\hbar}T_{\beta\alpha}(\hat{\mathbf{k}}_b, \hat{\mathbf{k}}_a)\frac{(\mu_a k_a)^{1/2}}{2\pi\hbar}. \quad (3.2.9)$$

Then

$$\frac{d\sigma_{\beta\alpha}}{d\Omega_b}(\hat{\mathbf{k}}_b, \hat{\mathbf{k}}_a) = \left(\frac{2\pi}{k_a}\right)^2|\delta_{\beta\alpha}\delta(\hat{\mathbf{k}}_b, \hat{\mathbf{k}}_a) - S_{\beta\alpha}(\hat{\mathbf{k}}_b, \hat{\mathbf{k}}_a)|^2. \quad (3.2.10)$$

The average cross-section for an unpolarized beam and target and any final spin orientation is

$$\frac{d\sigma_{ba}}{d\Omega_b}(\hat{\mathbf{k}}_b, \hat{\mathbf{k}}_a) = \frac{1}{(2s_1+1)(2s_2+1)}\sum_{s'm_s',sm_s}\frac{d\sigma_{\beta\alpha}}{d\Omega_b}(\hat{\mathbf{k}}_b, \hat{\mathbf{k}}_a). \quad (3.2.11)$$

The S-matrix is *unitary*:

$$\sum_\gamma\int d\Omega_c S^*_{\gamma\beta}(\hat{\mathbf{k}}_c, \hat{\mathbf{k}}_b)S_{\gamma\alpha}(\hat{\mathbf{k}}_c, \hat{\mathbf{k}}_a) = \delta_{\beta\alpha}\delta(\hat{\mathbf{k}}_b, \hat{\mathbf{k}}_a). \quad (3.2.12)$$

Now the 'partial wave expansion'

$$\begin{aligned} S_{\beta\alpha}(\hat{\mathbf{k}}_{b,\mathbf{k}_a}) &\equiv S_{bs'm_s',asm_s}(\hat{\mathbf{k}}_b, \hat{\mathbf{k}}_a) \\ &= \sum_{l'm',lm,jm_j}\langle l's'm'm_s' \mid jm_j\rangle\langle lsmm_s \mid jm_j\rangle S^j_{bl's',als}Y_{l'm'}(\hat{\mathbf{k}}_b)Y^*_{lm}(\hat{\mathbf{k}}_a) \end{aligned} \quad (3.2.13)$$

defines

$$S^j_{bl's',als} = S\text{-matrix in } (alsj) \text{ representation} \quad (3.2.14)$$

Here, j is the total spin: $\mathbf{j} = \mathbf{s} + \mathbf{l} = \mathbf{s}' - \mathbf{l}'$.
Properties:

unitarity (from flux conservation)

$$\sum_{cl''s''}(S^j_{cl''s'',bl's'})^*(S^j_{cl''s'',als}) = \delta_{ba}\delta_{l'l}\sigma_{s's}, \quad (3.2.15)$$

symmetry (from time reversal invariance)

$$S^j_{bl's',als} = S^j_{als,bl's'}. \quad (3.2.16)$$

The integrated cross-section becomes

$$\sigma_{ba} \equiv \int \mathrm{d}\Omega_b \frac{\mathrm{d}\sigma_{ba}}{\mathrm{d}\Omega_b}$$

$$= \frac{\pi}{k_a^2} \sum_{ll'_j} \frac{2j+1}{(2s_1+1)(2s_2+1)} |\delta_{ba}\delta_{l'l}\delta_{s's} - S^j_{bl's',als}|^2. \qquad (3.2.17)$$

Next we specialize to (i) scattering reactions of spinless particles and (ii) elastic scattering of spin-0 and spin-$\frac{1}{2}$ particles.

3.3 Scattering reactions of spinless particles

In this case $j = l$ and the partial-wave S-matrix (3.2.14) reduces to

$$S_{bl',al} = S_{ba,l}\delta_{l'l} \qquad (3.3.1)$$

and the partial-wave expansion becomes

$$S_{ba}(\hat{\mathbf{k}}_b, \hat{\mathbf{k}}_a) = S_{ba}(\hat{\mathbf{k}}_b \cdot \hat{\mathbf{k}}_a) = \frac{1}{2\pi} \sum_l \left(l + \frac{1}{2}\right) S_{ba,l} P_l(\cos\theta) \qquad (3.3.2)$$

where $\cos\theta = \hat{\mathbf{k}}_b \cdot \hat{\mathbf{k}}_a$.

The unitarity relation (3.2.15) becomes

$$\sum_c S^*_{cb,l} S_{ca,l} = \delta_{ba}. \qquad (3.3.3)$$

Consider now the diagonal elements

$$S_{a,al} \equiv S_l. \qquad (3.3.4)$$

Then eqn (3.3.3) for $b = a$ reads

$$1 = \sum_c |S_{ca,l}|^2 = |S_l|^2 + \sum_r |S_{ra,l}|^2 \qquad (3.3.5)$$

where $r \neq a$ runs over all nonelastic (reaction) channels.

The integrated cross-section (3.2.17) becomes (writing $k_a = k$)

$$\sigma_{ba} = \sum_l \sigma_{ba,l} \qquad \sigma_{ba,l} = \frac{2\pi}{k^2}(l + \tfrac{1}{2})\, |\delta_{ba} - S_{ba,l}|^2. \qquad (3.3.6)$$

The elastic dross-section is

$$\sigma_{\mathrm{E}} = \sigma_{aa} = \sum_i \sigma_{aa,l}$$

$$\sigma_{aa,l} = \sigma_{\mathrm{E},l} = \frac{2\pi}{h^2}(l + \tfrac{1}{2})\, |1 - S_l|^2. \qquad (3.3.7)$$

The total reaction cross-section is

$$\sigma_{\mathrm{R}} = \sum_r \sigma_{ra} = \sum_l \sigma_{\mathrm{R},l}$$

$$\sigma_{\mathrm{R},l} = \sum_r \sigma_{ra,l} = \frac{2\pi}{k^2}(l+\tfrac{1}{2})\sum_r |S_{ra,l}|^2. \tag{3.3.8}$$

The total cross-section is

$$\sigma_{\mathrm{T}} = \sigma_{\mathrm{E}} + \sigma_{\mathrm{R}} = \sum_c \sigma_{\mathrm{c}a} = \sum_l \sigma_{\mathrm{T},l}$$

$$\sigma_{\mathrm{T},l} = \frac{2\pi}{k^2}(l+\tfrac{1}{2})\left(|1-S_l|^2 + \sum_r |S_{ra,l}|^2\right) \tag{3.3.9}$$

and using elastic unitarity (3.3.5)

$$\sigma_{\mathrm{T},l} = \frac{2\pi}{k^2}(l+\tfrac{1}{2})(|1-S_l|^2 + 1 - |S_l|^2)$$

$$= \frac{4\pi}{k^2}(l+\tfrac{1}{2})(1-\mathrm{Re}\, S_l). \tag{3.3.10}$$

Consider now the elastic scattering amplitude (writing $\mathbf{k}_a = \mathbf{k}$, $\mathbf{k}_n = \mathbf{k}'$ in this case):

$$f(\theta) = f_{aa}(\hat{\mathbf{k}}' \cdot \hat{\mathbf{k}}) = \mathrm{i}\frac{2\pi}{k}\{\delta(\hat{\mathbf{k}}' \cdot \hat{\mathbf{k}}) - S_{aa}(\hat{\mathbf{k}}' \cdot \hat{\mathbf{k}})\}$$

$$= \frac{\mathrm{i}}{k}\sum_l (l+\tfrac{1}{2})(1-S_l)P_l(\cos\theta). \tag{3.3.11}$$

(partial wave expansion of the elastic scattering amplitude). Comparison of (3.3.10) and (3.3.11) yields the 'optical theorem'

$$\sigma_{\mathrm{T}} = \frac{4\pi}{k}\,\mathrm{Im}\, f(0). \tag{3.3.12}$$

3.4 Elastic scattering of spin-$\frac{1}{2}$ particles

Consider now the elastic scattering of spin-$\frac{1}{2}$ projectiles by spin-0 targets. Then the elastic scattering amplitude $f(\theta)$ becomes a 2×2 matrix in spin space:

$$f(\theta) = A(\theta)\times \mathbf{1} + B(\theta)(\mathbf{n}\cdot\boldsymbol{\sigma}), \tag{3.4.1}$$

where $\mathbf{n} = \hat{\mathbf{k}} \times \hat{\mathbf{k}}'$ is the unit vector normal to the scattering plane, $\boldsymbol{\sigma}$ is the Pauli spin vector and $\mathbf{1}$ denotes the 2×2 unit matrix. The amplitudes

$A(\theta)$ and $B(\theta)$ are known as the 'non-spin-flip' and 'spin-flip' amplitudes respectively.

The elastic S-matrix reduces to

$$S^j_{al'\frac{1}{2},al} = S^j_{l'l} = S^j_l \delta_{l'l}, \tag{3.4.2}$$

where $j = l \pm \frac{1}{2}$. With the shorter notation

$$S_l^{l+1/2} = S_l^+, \qquad S_l^{l-1/2} = S_l^- \tag{3.4.3}$$

the partial wave expansion of the non-spin-flip and spin-flip amplitudes becomes

$$\begin{aligned} A(\theta) &= \frac{\mathrm{i}}{2k} \sum_l \{(l+1)(1-S_l^+) + l(1-S_l^-)\} P_l(\cos\theta) \\ &= \frac{\mathrm{i}}{k} \sum_l (l+\tfrac{1}{2})(1-\bar{S}_l) P_l(\cos\theta) \end{aligned} \tag{3.4.4}$$

$$\begin{aligned} B(\theta) &= -\frac{1}{2k} \sum_l (S_l^+ - S_l^-) \frac{\mathrm{d}}{\mathrm{d}\theta} P_l(\cos\theta) \\ &= -\frac{1}{2k} \sum_l \xi_l \frac{\mathrm{d}}{\mathrm{d}\theta} P_l(\cos\theta), \end{aligned} \tag{3.4.5}$$

where we have defined the average and the difference of the elastic S-matrix elements in the two spin states

$$\bar{S}_l = \frac{l+1}{2l+1} S_l^+ + \frac{l}{2l+1} S_l^-, \qquad \xi_l = S_l^+ - S_l^-. \tag{3.4.6}$$

The differential elastic scattering cross-section is

$$\frac{\mathrm{d}\sigma}{\mathrm{d}\Omega} = |A(\theta)|^2 + |B(\theta)|^2. \tag{3.4.7}$$

Scattering in general changes the state of polarization of the particles. We confine ourselves to an unpolarized incoming beam. The polarization vector is $\mathbf{P} = P(\theta)\mathbf{n}$ with the polarization given by

$$P(\theta) = 2\,\frac{\operatorname{Im} A(\theta) \operatorname{Im} B(\theta) + \operatorname{Re} A(\theta) \operatorname{Re} B(\theta)}{\mathrm{d}\sigma/\mathrm{d}\Omega} \tag{3.4.8}$$

with $\mathrm{d}\sigma/\mathrm{d}\Omega$ given by eqn (3.4.7).

The integrated cross-sections become

$$\sigma_{\mathrm{E}} = \frac{2\pi}{k^2} \sum_l (l+\tfrac{1}{2}) \left\{ |1-\bar{S}_l|^2 + \frac{l(l+1)}{(2l+1)^2} |\xi_l|^2 \right\} \tag{3.4.9}$$

$$\sigma_{\mathrm{R}} = \frac{2\pi}{k^2}\sum_l (l+\tfrac{1}{2})\left\{1-|\bar{S}_l|^2-\frac{l(l+1)}{(2l+1)^2}|\xi_l|^2\right\} \tag{3.4.10}$$

$$\sigma_{\mathrm{T}} = \frac{4\pi}{k^2}\sum_l (l+\tfrac{1}{2})(1-R_l\bar{S}_l). \tag{3.4.11}$$

It follows that the optical theorem still holds in the form (3.3.12).

3.5 Coulomb effects

In the collision of charged particles it is convenient to take out the contribution from Coulomb (Rutherford) scattering explicitly. We define the reduced S-matrix S by writing (3.2.14) in the form

$$S^j_{bl's',als} = \exp\{\mathrm{i}\sigma_{l'}(b)\}S^j_{bl's',als}\exp\{\mathrm{i}\sigma_l(a)\}. \tag{3.5.1}$$

Here $\sigma_l(c)$ are the phase shifts for Rutherford scattering in channel c given by

$$\exp\{\mathrm{i}2\sigma_l(c)\} = \frac{\Gamma(l+1+\mathrm{i}n_c)}{\Gamma(l+1-\mathrm{i}n_c)}, \tag{3.5.2}$$

where n_c is the Sommerfeld parameter in channel c, defined in terms of the charges Z_1e and Z_2e of the nuclei and their relative asymptotic velocity v_c in that channel

$$n_c = \frac{Z_1Z_2e^2}{\hbar v_c}. \tag{3.5.3}$$

With the definition (3.2.14) we can write

$$S^j_{bl's',als} = \delta_{ba}\delta_{l'l}\delta_{s's}\exp\{\mathrm{i}2\sigma_l(a)\} + \exp\{\mathrm{i}\sigma_{l'}(b)\}(S^j_{bl's',als}-\delta_{ba}\delta_{l'l}\delta_{s's})\exp\{\mathrm{i}\sigma_l(a)\}. \tag{3.5.4}$$

The first term then gives rise to the Rutherford scattering amplitude

$$f_c(\theta) = \frac{\mathrm{i}}{k}\sum_l (l+\tfrac{1}{2})\{1-\exp(\mathrm{i}2\sigma_l)\}P_l(\cos\theta)$$
$$= -\frac{n}{2k}\frac{1}{(\sin\frac{1}{2}\theta)^2}\exp[\mathrm{i}\{2\sigma_0-2n\ln(\sin\tfrac{1}{2}\theta)\}] \tag{3.5.5}$$

where we have written $n = n_a$ and $\sigma_l = \sigma_l(a)$.

For elastic scattering of spinless particles the amplitude becomes

$$f(\theta) = f_c(\theta) + \frac{\mathrm{i}}{k}\sum_l (l+\tfrac{1}{2})(1-s_l)\exp(\mathrm{i}2\sigma_l)P_l(\cos\theta) \tag{3.5.6}$$

where

$$s_l = \eta_l \exp(\mathrm{i}2\nu_l) \quad \text{with} \quad \nu_l = \delta_l - \sigma_l. \tag{3.5.7}$$

In considering the application of the formalism of Sections 3.1–3.4 to collisions of charged particles it can be assumed that the long-range Coulomb forces are effectively screened at large distances from the nuclear or atomic scattering centres (for a detailed discussion see Goldberger and Watson (1964) [3.1] who refer to earlier work on this subject).

3.6 Collisions involving identical particles

Suppose that in a collision $\alpha \to \beta$ the two particles in the initial channel or in the final channel or in both are identical. Then the scattering amplitude has to be symmetrized with respect to exchange of the identical particles. We denote the channel spin in the initial channel by $s = S_a = 0, 1, 2, \ldots, 2I_a$ and in the final channel by $s' = S_b = 0, 1, 2, \ldots, 2I_b$, where $s_1 = s_2 = I_a$ and $s'_1 = s'_2 = I_b$ are the corresponding intrinsic spins.

We distinguish four cases I-IV according to the scheme

		initial channel α	
		non-identical	identical
final channel β	non-identical	I	II
	identical	III	IV

The symmetrized scattering amplitudes $\tilde{f}_{\beta\alpha}(\hat{\mathbf{k}}_b, \hat{\mathbf{k}}_a)$ are then given by (see Goldberger and Watson (1964) [3.1])

$$\text{(I)} \quad \tilde{f}_{\beta\alpha}(\hat{\mathbf{k}}_b, \hat{\mathbf{k}}_a) = f_{\beta\alpha}(\hat{\mathbf{k}}_b, \hat{\mathbf{k}}_a)$$

$$\text{(II)} \quad \tilde{f}_{\beta\alpha}(\hat{\mathbf{k}}_b, \hat{\mathbf{k}}_a) = \frac{1}{\sqrt{2}}\{f_{\beta\alpha}(\hat{\mathbf{k}}_b, \hat{\mathbf{k}}_a) + \tau_a(-1)^{2I_a - S_a} f_{\beta\alpha}(\hat{\mathbf{k}}_b, -\hat{\mathbf{k}}_a)\}$$

$$\text{(III)} \quad \tilde{f}_{\beta\alpha}(\hat{\mathbf{k}}_b, \hat{\mathbf{k}}_a) = \frac{1}{\sqrt{2}}\{f_{\beta\alpha}(\hat{\mathbf{k}}_b, \hat{\mathbf{k}}_a) + \tau_b(-1)^{2I_b - S_b} f_{\beta\alpha}(-\hat{\mathbf{k}}_b, \hat{\mathbf{k}}_a)\} \tag{3.6.1}$$

$$\begin{aligned}\text{(IV)} \quad \tilde{f}_{\beta\alpha}(\hat{\mathbf{k}}_b, \hat{\mathbf{k}}_a) = \tfrac{1}{2}\{&f_{\beta\alpha}(\hat{\mathbf{k}}_b, \hat{\mathbf{k}}_a) + \tau_a(-1)^{2I_a - S_a} f_{\beta\alpha}(\hat{\mathbf{k}}_b, -\hat{\mathbf{k}}_a) \\ &+ \tau_b(-1)^{2I_b - S_b} f_{\beta\alpha}(-\hat{\mathbf{k}}_b, \hat{\mathbf{k}}_a) \\ &+ \tau_a\tau_b(-1)^{2(I_a + I_b) - (S_a + S_b)} f_{\beta\alpha}(-\hat{\mathbf{k}}_b, -\hat{\mathbf{k}}_a)\}\end{aligned}$$

where

$$\tau_c = \begin{matrix} +1 & \text{for bosons} \\ -1 & \text{for fermions} \end{matrix} \quad \text{in channel } c.$$

For elastic scattering, where

$$f_{\alpha\alpha}(\hat{\mathbf{k}}',\hat{\mathbf{k}})=f_{\alpha\alpha}(-\hat{\mathbf{k}}',-\hat{\mathbf{k}})\equiv f_s(\theta)$$
$$F_{\alpha\alpha}(\hat{\mathbf{k}}',\hat{\mathbf{k}})=f_{\alpha\alpha}(-\hat{\mathbf{k}}',\hat{\mathbf{k}})\equiv f_s(\pi-\theta), \tag{3.6.2}$$

the symmetrized amplitude for a given channel spin S,

$$\tilde{f}_s(\theta)=f_s(\theta)+\tau(-1)^{2I-s}f_s(\pi-\theta), \tag{3.6.3}$$

gives the differential cross-section

$$\sigma_s(\theta)=|\tilde{f}_s(\theta)|^2=|f_s(\theta)|^2+|f_s(\pi-\theta)|^2+\tau(-1)^{2I-S}2\mathrm{Re}\{f_s(\theta)f_s^*(\pi-\theta)\}. \tag{3.6.4}$$

For random initial spin orientations we have

$$\sigma(\theta)=\frac{1}{(2I+1)^2}\sum_{s=0}^{2I}(2S+1)\sigma_s(\theta). \tag{3.6.5}$$

If the interaction is spin independent, $f_s=f$; this yields

$$\sigma(\theta)=|f(\theta)|^2+|f(\pi-\theta)|^2+\frac{\tau}{2I+1}2\mathrm{Re}\{f(\theta)f^*(\pi-\theta)\}. \tag{3.6.6}$$

All angular distributions for collisions involving identical particles are symmetrical about $\theta=90°$ in the c.m. system.

From the partial-wave expansion (3.2.13) we obtain, because of $Y_{lm}(-\hat{\mathbf{k}})=(-1)^lY_{lm}(\hat{\mathbf{k}})$, for the symmetrized S-matrix

$$\tilde{S}^{j(s)}_{bl',nl}-\delta_{ba}\delta_{l'l}=\zeta(S^{j(s)}_{bl',nl}-\delta_{ba}\delta_{l'l}), \tag{3.6.7}$$

where

(I) $$\zeta=1$$

(II) $$\zeta=\frac{1}{\sqrt{2}}\{1+\tau_a\,\mathrm{sgn}(2I_a-S_a+l)\}$$

(III) $$\zeta=\frac{1}{\sqrt{2}}\{1+\tau_b\,\mathrm{sgn}(2I_b-S_b+l')\} \tag{3.6.8}$$

(IV) $$\zeta=\tfrac{1}{2}\{1+\tau_a\,\mathrm{sgn}(2I_a-S_a+l)+\tau_b\,\mathrm{sgn}(2I_b-S_b+l')+\tau_a\tau_b\,\mathrm{sgn}(2I_a+2I_b-S_a-S_b+l+l')\}.$$

In particular, for elastic scattering we have for case (IV) with $b=a$,

$$\tilde{S}^{j(s)}_{l'l}-\delta_{l'l}=\zeta(S^{j(s)}_{l'l}-\sigma_{l'l}) \tag{3.6.9}$$

with

$$\zeta=\tfrac{1}{2}\{1+(-1)^{l+l'}\}+\tfrac{1}{2}\{(-1)^l+(-1)^{l'}\}\tau(-1)^{2I-s}. \tag{3.6.10}$$

Thus the effect of identity on the partial-wave S-matrix is to suppress either even- or odd-l partial waves, depending on the intrinsic spin and statistics of the colliding nuclei.

3.7 Elastic unitarity and absorption

The diagonal elements S_l of the partial-wave S matrix for elastic scattering can be regarded as a (complex-valued) function of orbital angular momentum l and wave number k or energy E: the 'scattering function'. Elastic unitarity (3.3.5) implies

$$|S_l|^2 - 1 - \sum_r |S_{ra,l}|^2 \leqslant 1, \tag{3.7.1}$$

where the equality sign for a given l holds only if all $S_{ra,l} = 0$. If we write S_l in the form

$$S_l = \eta_l \exp(\mathrm{i}2\delta_l) \tag{3.7.2}$$

where η_l and the elastic phase shifts δ_l are real numbers, we have, using (3.3.8),

$$\eta_l^2 = 1 - \frac{k^2 \sigma_{\mathrm{R},l}}{2\pi(l+\frac{1}{2})} \leqslant 1. \tag{3.7.3}$$

For projectiles of spin $\frac{1}{2}$ we define

$$S_l^{\pm} = \eta_l^{\pm} \exp(\mathrm{i}2\delta_l^{\pm}). \tag{3.7.4}$$

From elastic unitarity

$$\sum_c |S^j_{ca,l}|^2 = 1 \qquad \text{for } j = l \pm \tfrac{1}{2} \tag{3.7.5}$$

we have

$$(\eta_l^{\pm})^2 = 1 - \sum_r |S^{\pm}_{ra,l}|^2 = 1 - \frac{k^2 \sigma^{\pm}_{\mathrm{R},l}}{2\pi(l+\frac{1}{2})} \leqslant 1. \tag{3.7.6}$$

Equations (3.7.3) and (3.7.6) show that the presence of open nonelastic channels such that $\sigma_{\mathrm{R},l} \neq 0$ causes a depletion of flux in the elastic channel which is called 'absorption'. A given partial wave l is called 'strongly absorbed' or 'weakly absorbed' according to whether

$$\eta_l \ll 1 \quad \text{or} \quad 1 - \eta_l \ll 1. \tag{3.7.7}$$

From the partial-wave expansion of the elastic scattering amplitude,

$$f(\theta) = \frac{\mathrm{i}}{k} \sum_l (l+\tfrac{1}{2})\{1 - \eta_l \exp(\mathrm{i}2\delta_l)\} P_l(\cos\theta), \tag{3.7.8}$$

it can be seen that, if a 'band' of several adjacent partial waves (of width Δl in l-space) is strongly absorbed, the amplitude $f(\theta)$ and hence the differential cross-section will exhibit a quantal diffraction pattern (of width $\Delta\theta \approx 1/\Delta l$) according to the general conditions for quantum diffraction described in Chapter 1. Thus the presence of strong absorption, which is usually due to many open reaction channels in nuclear collisions, is the basis for the occurrence of diffraction effects in nuclear scattering.

By an optical analogy the interaction region that corresponds to the band of strongly absorbed partial waves can be regarded as an opaque 'amplitude object' limiting or restricting the amplitude distribution of the incoming plane wave in l-space. The resultant elastic scattering can then be regarded as 'shadow scattering' caused by all the nonelastic processes. However, the presence of the real phase shifts δ_l means that the nuclear interaction represents in general a 'phase object' as well as an 'amplitude object'. Roughly speaking†, η_l describes the absorptive properties and δ_l the refractive properties of the interaction.

3.8 Dynamical theories

Having discussed the *asymptotic* description of collisions in the stationary formulation, we now turn to the question of how the scattering amplitudes, and hence the S-matrix, are determined by the dynamics of the collision process. In our case the colliding particles are atomic nuclei whose internal structure may change during the collision.

Consider again the process described by (3.1.1). The internal states of the particles in channel α, denoted by ϕ_α, satisfy the Schrödinger equation

$$h_\alpha \phi_\alpha = \varepsilon_\alpha \phi_\alpha \tag{3.8.1}$$

where

$$h_\alpha = h_{\alpha_1} + h_{\alpha_2}, \qquad \varepsilon_\alpha = \varepsilon_{\alpha_1} + \varepsilon_{\alpha_2}. \tag{3.8.2}$$

The total wave function in channel α, Ψ_α, satisfies the Schrödinger equation

$$H\Psi = (\varepsilon_\alpha + E_a)\Psi_\alpha, \tag{3.8.3}$$

where

$$H = h_\alpha + K_a + U_\alpha. \tag{3.8.4}$$

† It should be remembered that absorptive interactions in wave mechanics, for instance as described by the imaginary part of an 'optical potential', affect the phase shifts δ_l as well as the η_l.

Here K_a is the kinetic energy operator of the relative motion with eigenvalues $E_a = \hbar^2 k_a^2 / 2\mu_a$ and U_α is the potential energy operator of interaction between nuclei a_1 and a_2. Similar equations hold for the wave functions in channels β, γ, etc., with the same total energy in all channels: $E_\alpha + E_a = \varepsilon_\beta + E_b$.

The asymptotic form of the wave function Ψ_α, with boundary conditions corresponding to an incoming plane wave in the entrance channel α and outgoing spherical waves in all exit channels β, is given by eqn (3.2.2). Our aim is to express the scattering amplitude $f_{\beta\alpha}$ or the transition matrix $T_{\beta\alpha}$ in terms of the interaction operator U. To this end we expand the total wave function $\Psi^{(+)}$ in a complete set of internal eigenstates ϕ_γ:

$$\Psi^{(+)} = \sum_\gamma \psi_\gamma^{(+)} \phi_\gamma. \tag{3.8.5}$$

The expansion coefficients, obtained by scalar multiplication with $\phi_\beta = \phi_\beta(\xi_\beta)$ which involves integration over the internal variables ξ_β,

$$\psi_\beta^{(+)} = \langle \phi_\beta \mid \psi^{(+)} \rangle \tag{3.8.6}$$

are functions of the relative coordinate $\mathbf{r}_b$. By inserting (3.8.5) in (3.8.3) and performing scalar multiplication with ϕ_β we obtain

$$(K_b - E_b)\psi_\beta^{(+)} = -\sum_\gamma \langle \phi_\beta \,|U_\beta|\, \phi_\gamma \rangle \psi_\gamma^{(+)}, \tag{3.8.7}$$

which is a system of coupled equations for the expansion coefficients $\psi^{(+)}$. By means of the free-space Green function for outgoing-wave boundary conditions, which satisfies

$$-(\nabla_b^2 + k_b^2) G_0^{(+)}(\mathbf{r}_b - \mathbf{r}_b') = \delta(\mathbf{r}_b - \mathbf{r}_b') \tag{3.8.8}$$

and has the form

$$G_0^{(+)}(\mathbf{r}_b, \mathbf{r}_b') = -\frac{1}{4\pi} \frac{\exp(\mathrm{i}k_b\,|\mathbf{r}_b - \mathbf{r}_b'|)}{|\mathbf{r}_b - \mathbf{r}_b'|}, \tag{3.8.9}$$

we can integrate (3.8.7) as

$$\psi_\beta^{(+)} = \exp(\mathrm{i}\mathbf{k}_a \cdot \mathbf{r}_a)\delta_{\beta\alpha} + \frac{2\mu_b}{\hbar^2} \int \mathrm{d}^3 r_b' G_0^{(+)}(\mathbf{r}_b, \mathbf{r}_b') \sum_\gamma \langle \phi_\beta \,|U_\beta|\, \phi_\gamma \rangle \psi_\gamma^{(+)}. \tag{3.8.10}$$

Our choice of the first right-hand term in (3.8.10) makes this particular solution of eqn (3.8.7) the expansion coefficient of the total wave function $\psi_\alpha^{(+)} = \sum_\beta \psi_\beta^{(+)} \phi_\beta$ in the entrance channel. Its asymptotic form is obtained by using the asymptotic expansion of the Green function for $r_b \gg r'_b$,

$$G_0^{(+)}(\mathbf{r}_b, \mathbf{r}'_b) \approx -\frac{1}{4\pi} \exp(-i\mathbf{r}_b \cdot \mathbf{r}'_b) \frac{\exp(ik_b r_b)}{r_b} \tag{3.8.11}$$

where $\mathbf{k}_b = \mathbf{r}_b \hat{\mathbf{r}}_b$, in eqn (3.8.10). This yields

$$\Psi_\alpha^{(+)} \approx \exp(i\mathbf{k}_a \cdot \mathbf{r}_a)\phi_\alpha - \sum_\beta \frac{\mu_b}{2\pi\hbar^2} \int d^3r'_b \langle \exp(i\mathbf{k}_b \cdot \mathbf{r}'_b)\phi_\beta \,|U_\beta|\, \Psi_\alpha^{(+)} \rangle \phi_\beta \frac{\exp(i\mathbf{k}_b \mathbf{r}_b)}{r_b} \tag{3.8.12}$$

Comparison with eqn (3.2.8) yields the T-matrix elements,

$$T_{\beta\alpha} = \langle \exp(i\mathbf{k}_b \cdot \mathbf{r}'_b)\phi_\beta \,|U_\beta|\, \Psi_\alpha^{(+)} \rangle. \tag{3.8.13}$$

The interaction operator U_β depends in general on all internal coordinates ξ_β, spatial as well as spin, of the nuclei b_1 and b_2 in channel β, on the relative coordinate $\mathbf{r}_b$ and the channel spin $\mathbf{s}'$, and on the charge numbers Z_{b_1} and Z_{b_2}. Thus it may cause a variety of processes which are categorized as follows:

(1) *elastic scattering*, in which the nuclei undergo no change in their internal energies, although there may be a change of spin orientation;
(2) *inelastic scattering*, in which a change of internal energies does occur but there is no change in composition;
(3) *rearrangement collision*, in which the colliding nuclei suffer a change in composition by exchange of one or more of their constituent nucleons.

Elastic scattering is not restricted to a mere deflection of the colliding nuclei with or without spin reorientation. The same type of nuclei may re-emerge with the same internal energies after having undergone a succession of changes in energy and composition. Thus nonelastic processes may contribute (in 'higher order') to elastic scattering. In certain instances it is appropriate to define elastic scattering more generally as a process in which the final and initial states are indistinguishable, even though the nuclei have undergone a physical rearrangement. This is the case in 'elastic transfer' reactions.

A very useful, if somewhat vague, distinction of *reaction mechanisms* is that between processes involving formation of a compound nucleus (CN) and those proceeding by direct interaction (DI). A thorough discussion of these modes of reaction, their physical nature and distinctive features, and the conditions under which the distinction is valid can be found in Austern (1970) [3.4]. A brief summary can be given as follows. The CN mechanism involves excitations of many degrees of freedom of the nuclear many-body surplus, while in DI processes only a few degrees of freedom are excited. In a time-dependent description, the duration of direct reactions is short, of the order of the transit time through the nucleus ($\sim 10^{-22}$ s), while CN reactions are several orders of magnitude slower. This difference manifests itself in a smooth energy dependence of cross-sections over intervals of several MeV for DI processes, in contrast with rapid variations of CN cross-sections over ranges of the order of keV or less. A more basic definition can be given in terms of the 'closed' and 'open' parts of a stationary-state wave function Ψ. The open part of Ψ includes all those contributions which change at asymptotic distances and thus represent incoming and outgoing particles, while the closed part of Ψ consists of all contributions that go to zero in the asymptotic region of configuration space. A DI mechanism is now defined as one that involves the coupling of only the various contributions of the open part of Ψ among each other, while a CN mechanism involves the coupling between open and closed contributions to the total wave function.

For DI, the interaction operator U_β depends on $\mathbf{r}_b$ and on only a few of the internal coordinates ξ_β. Now we split U_β into two parts:

$$U_\beta = U_b + U_\beta, \tag{3.8.14}$$

where U_b is an average interaction independent of the ξ_β, which is called the 'optical potential'. It is a function of $\mathbf{r}_b$ only, but may also depend on the channel spin $\mathbf{s}'$. For charged particles, U_b includes the Coulomb interaction $V_{c,b}(\mathbf{r}_b)$,

$$U_b(\mathbf{r}_b) = V_{c,b}(r_b) + V_{N,b}(\mathbf{r}_b). \tag{3.8.15}$$

With (3.8.14), eqn (3.8.7) can be written in the form

$$(K_b + U_b - E_b)\psi_\beta^{(+)} = -\sum_\gamma \langle\phi_\beta\,|U_\beta|\,\phi_\gamma\rangle\psi_\gamma^{(+)}. \tag{3.8.16}$$

By means of the Green function $G^{(+)}(\mathbf{r}_b, \mathbf{r}_b')$ that satisfies

$$\frac{2\mu_b}{\hbar^2}(K_b + U_b - E_b)G^{(+)}(\mathbf{r}_b, \mathbf{r}_b') = \delta(\mathbf{r}_n - \mathbf{r}_b') \tag{3.8.17}$$

we can integrate (3.8.16) as

$$\psi_\beta^{(+)} = \chi_\alpha^{(+)}\delta_{\beta\alpha} + \frac{2\mu_b}{\hbar^2}\int d^3r_b' G^{(+)}(\mathbf{r}_b, \mathbf{r}_b')\sum_\gamma \langle\phi_\beta\,|U_\beta|\,\phi_\gamma\rangle\psi_\gamma^{(+)} \tag{3.8.18}$$

Here, $\chi_\alpha^{(+)} = \chi_\alpha^{(+)}(\mathbf{k}_a, \mathbf{r}_a)$ is the wave function for elastic scattering by the optical potential U_a, i.e. a solution of the Schrödinger equation

$$(K_a + U_a - E_a)\chi_\alpha^{(+)} = 0 \tag{3.8.19}$$

which for neutral particles behaves asymptotically as

$$X_\alpha(\mathbf{k}_a, \mathbf{r}_a) \simeq \exp(\mathrm{i}\mathbf{k}_a \cdot \mathbf{r}_a) + f(\hat{\mathbf{k}}'_a, \hat{\mathbf{k}}_a)\frac{\exp(\mathrm{i}k_a r_a)}{r_a}. \tag{3.8.20}$$

The Green function $G^{(+)}$ has the asymptotic form

$$G^{(+)}(\mathbf{r}_b, \mathbf{r}'_b) \simeq -\frac{1}{4\pi}\chi_\beta^{(-)*}(\mathbf{k}_b, \mathbf{r}'_b)\frac{\exp(\mathrm{i}k_b r_b)}{r_b}, \tag{3.8.21}$$

where $\chi^{(-)}$ is the time-reversed scattering wave function which is related to $\chi^{(+)}$ by the Wigner relation

$$x^{(-)*}(\mathbf{k}, \mathbf{r}) = x^{(+)}(-\mathbf{k}, \mathbf{r}). \tag{3.8.22}$$

Thus the asymptotic form of $\psi_\beta^{(+)}$ becomes

$$\begin{aligned}\psi_\beta^{(+)} \simeq {}& \exp(\mathrm{i}\mathbf{k}_a\mathbf{r}_a)\delta_{\beta\alpha} + f(\hat{\mathbf{k}}'_a, \hat{\mathbf{k}}_n)\frac{\exp(\mathrm{i}k_a r_a)}{r_a}\delta_{\beta\alpha} \\ & -\frac{\mu_b}{2\pi\hbar^2}\int \mathrm{d}^3 r'_b \chi_\beta^{(-)*}(\mathbf{k}_b, \mathbf{r}'_b)\sum_\gamma \langle\phi_\beta\,|U_\beta|\,\phi_\gamma\rangle\psi_\gamma^{(+)}\frac{\exp(\mathrm{i}k_b r_b)}{r_b}.\end{aligned} \tag{3.8.23}$$

Hence

$$T_{\beta\alpha} = -\frac{2\pi\hbar^2}{\mu_\mathrm{a}} f(\hat{\mathbf{k}}_a, \hat{\mathbf{k}}_a)\delta_{\beta\alpha} + \langle\chi_\beta^{(-)}\phi_\beta\,|U_\beta|\,\Psi_\alpha^{(+)}\rangle. \tag{3.8.24}$$

By introducing the T-matrix for elastic scattering by the potential U_b, which is obtained from eqn (3.8.13) on replacing U_β by U_b and $\Psi_\alpha^{(+)}$ by $\chi_\alpha^{(+)}\phi_\alpha$

$$\begin{aligned}T_\mathrm{E}\delta_{\beta\alpha} &= \langle\exp(\mathrm{i}\mathbf{k}_b \cdot \mathbf{r}'_b)\phi_\beta\,|U_b|\,\varepsilon_\alpha\chi_\alpha^{(+)}\rangle \\ &= \langle\exp(\mathrm{i}\mathbf{k}_a \cdot \mathbf{r}'_a)\,|U_\mathrm{a}|\,\chi_\alpha^{(+)}(\mathbf{k}_a, \mathbf{r}_a)\rangle\delta_{\beta\alpha} \\ &= -\frac{2\pi\hbar^2}{\mu_a} f(\hat{\mathbf{k}}'_a, \hat{\mathbf{k}}_a)\delta_{\beta\alpha},\end{aligned} \tag{3.8.25}$$

we can write

$$T_{\beta\alpha} = T_\mathrm{E}\delta_{\beta\alpha} + T_{\mathrm{DI},\beta\alpha} \tag{3.8.26}$$

or

$$\begin{aligned}\langle\exp(\mathrm{i}\mathbf{k}_b \cdot \mathbf{r}'_b)\phi_\beta\,|U_\beta|\,\Psi_\alpha^{(+)}\rangle = {}& \langle\exp(\mathrm{i}\mathbf{k}_b \cdot \mathbf{r}_b)\phi_\beta\,|U_b|\,\phi_\alpha\chi_\alpha^{(+)}\rangle \\ & + \langle\chi_\beta^{(-)}\phi_\beta\,|U_\beta|\,\Psi_\alpha^{(+)}\rangle.\end{aligned} \tag{3.8.27}$$

Equation (3.8.27) is known as the Gell-Mann–Goldberger (GM–G) relation.

3.9 References

[3.1] Goldberger, M. L. and Watson, K. M. (1964). *Collision theory*. John Wiley, New York.
[3.2] Messiah, A. (1962). *Quantum mechanics*. North-Holland, Amsterdam.
[3.3] Gottfried, K. (1966). *Quantum mechanics*. Benjamin, New York.
[3.4] Austern, N. (1970). *Direct nuclear reaction theories*. John Wiley, New York.

4

DIFFRACTION IN CONFIGURATION SPACE

4.1 Huygens' principle

It is convenient and constructive to begin with a treatment of diffraction in configuration space, because it is the precise quantal analogue of the diffraction theory for classical scalar wave fields and defines most of the concepts and conditions we shall use in other approaches. Consider again elastic scattering by a (complex) local potential $U(r)$ described by the Schrödinger equation

$$(\nabla^2 + k^2)\psi(\mathbf{r}) = \frac{2\mu}{\hbar^2} U(\mathbf{r})\psi(\mathbf{r}). \tag{4.1.1}$$

As we have seen in Section 3.2 this can be converted into the integral equation

$$\psi(\mathbf{r}) = \psi_i(\mathbf{r}) + \psi_s(\mathbf{r}) \tag{4.1.2}$$

where $\psi_i(\mathbf{r})$ is an incoming free wave and $\psi_s(\mathbf{r})$ is the scattered wave

$$\psi_s(\mathbf{r}) = \frac{2\mu}{\hbar^2} \int d^3r' G(\mathbf{r}, \mathbf{r}') U(\mathbf{r}') \psi(\mathbf{r}') \tag{4.1.3}$$

obtained using the free-space Green function defined by eqn (3.8.9).

Now let us assume that the interaction is confined to a domain Ω outside which $U(\mathbf{r})$ is exactly zero. In this outside region $\bar{\Omega}$, $\psi_s(\mathbf{r}')$ satisfies the free-wave equation. On multiplying this equation by $G(\mathbf{r}, \mathbf{r}')$ and integrating over $\bar{\Omega}$ we obtain

$$\int_{\bar{\Omega}} d^3r' G(\mathbf{r}, \mathbf{r}')(\nabla'^2 + k^2)\psi_s(\mathbf{r}') = 0. \tag{4.1.4}$$

On the other hand, from eqn (3.8.8) for $G(\mathbf{r}, \mathbf{r}')$ we find after multiplication by $\psi_s(\mathbf{r}')$ and integration over $\bar{\Omega}$,

$$\int_{\bar{\Omega}} d^3r' \psi_s(\mathbf{r}')(\nabla'^2 + k^2) G(\mathbf{r}, \mathbf{r}') = \int_{\bar{\Omega}} d^3r' \psi_s(\mathbf{r}')\, \delta(\mathbf{r} - \mathbf{r}') = \psi_s(\mathbf{r}) \tag{4.1.5}$$

for $\mathbf{r} \in \bar{\Omega}$. Subtraction of eqn (4.4) from (4.5) gives

$$\psi_s(\mathbf{r}) = \int_{\bar{\Omega}} d^3r' \{\psi_s(\mathbf{r}')\, \nabla'^2 G(\mathbf{r}, \mathbf{r}') - G(\mathbf{r}, \mathbf{r}')\, \nabla'^2 \psi_s(\mathbf{r}')\}. \tag{4.1.6}$$

The volume integral over $\bar{\Omega}$ can be converted into an integral over the surface Σ enclosing $\bar{\Omega}$ (or Ω) by means of Green's theorem with the result

$$\psi_s(\mathbf{r}) = \int_\Sigma d^2\rho\, \mathbf{n}_\rho \mathbf{\Gamma}(\mathbf{r}, \boldsymbol{\rho}), \tag{4.1.7}$$

where $\mathbf{\Gamma}$ is the 'Green vector' defined by

$$\mathbf{\Gamma}(\mathbf{r}, \boldsymbol{\rho}) = \psi_s(\boldsymbol{\rho})\, \boldsymbol{\nabla}_\rho G(\mathbf{r}, \boldsymbol{\rho}) - G(\mathbf{r}, \boldsymbol{\rho})\, \boldsymbol{\nabla}_\rho \psi_s(\boldsymbol{\rho}) \tag{4.1.8}$$

and $\mathbf{n}_\rho$ is the unit vector of the outward normal to Σ (i.e. pointing into $\bar{\Omega}$) at the point $\boldsymbol{\rho}$. Equation (4.1.7) is exact and represents the quantal analogue of *Huygens' principle:* the scattered field $\psi_s(\mathbf{r})$ at a point $\mathbf{r} \in \bar{\Omega}$ is a superposition of elementary spherical waves originating from the surface Σ that encloses the interaction domain

4.2 Boundary value and saltus problems

In eqn (4.1.7) the potential $U(\mathbf{r})$ no longer appears explicitly; instead we have to know the values $\psi_s(\boldsymbol{\rho})$ and $\boldsymbol{\nabla}_\rho \psi_s(\boldsymbol{\rho})$ on the surface Σ. These can be determined by solving the Schrödinger equation (4.1.1) in the interior domain Ω. However, if this is too complicated we may attempt to characterize the interaction by *prescribing* the values of ψ_s and its usual derivative on the surface Σ. Thereby we replace the potential scattering problem by a *boundary value problem.*

However, the functions $\psi_s(\boldsymbol{\rho})$ and $\nabla_\rho \psi_s(\boldsymbol{\rho})$ on Σ cannot both be independently prescribed; the solution $\psi_s(\mathbf{r})$ is uniquely determined by either of them. In fact the two functions are related by an integral equation (see, for example, Hönl *et al.* (1961 [4.1]) for a careful discussion of these questions in optical diffraction theory). If, nevertheless, we *do* prescribe both ψ_s and $|\boldsymbol{\nabla}\psi_s|$ on Σ, the expression (4.1.7) is still a solution of the wave equation, but the quantities $\psi_s \equiv \sigma(\boldsymbol{\rho})$ and $|\boldsymbol{\nabla}\psi_s| \equiv t(\boldsymbol{\rho})$ are in general no longer the boundary values that $\psi_s(\mathbf{r})$ and $|\boldsymbol{\nabla}\psi_s(\mathbf{r})|$ take on as $\mathbf{r}$ tends to $\boldsymbol{\rho}$. If we denote the right-hand side of eqn (4.1.7) by $\gamma(\mathbf{r})$, the quantities $\sigma(\boldsymbol{\rho})$ and $t(\boldsymbol{\rho})$ are the *discontinuities*

$$\begin{aligned} \sigma(\boldsymbol{\rho}) &= \gamma(\boldsymbol{\rho}+) - \gamma(\boldsymbol{\rho}-) \\ t(\boldsymbol{\rho}) &= |\boldsymbol{\nabla}\gamma|_{\rho+} - |\boldsymbol{\nabla}\gamma|_{\rho-} \end{aligned} \tag{4.2.1}$$

of γ and $|\nabla\gamma|$ on going across the surface Σ from the outside (+) to the inside (−). This procedure therefore amounts to replacing the boundary value problem by a 'saltus problem' and is the starting point of Kirchhoff's approximation which will be discussed in the next section.

4.3 Kirchhoff's approximation

In nuclear collisions the source of particles is practically at infinity and the incoming wave is a plane wave, $\psi_i(\mathbf{r})=\exp(i\mathbf{k}\cdot\mathbf{r})$. Then we can divide the enclosing surface into two parts, $\Sigma=J+D$, where J is the area that would be 'illuminated' by $\psi_i(\mathbf{r})$ according to geometrical optics and D is the complementary 'dark' portion of Σ (Fig. 4.1). Now we suppose the interaction is such that the enclosed region Ω can be regarded as perfectly absorbing. Thus Kirchhoff's approximation consists of postulating that

$$\begin{aligned} &\sigma(\boldsymbol{\rho})=\psi_i(\boldsymbol{\rho}), && t(\boldsymbol{\rho})=|\boldsymbol{\nabla}_\rho\psi_i(\boldsymbol{\rho})|, && \text{for } \boldsymbol{\rho} \text{ on } J \\ &\sigma(\boldsymbol{\rho})=0, && t(\boldsymbol{\rho})=0 && \text{for } \boldsymbol{\rho} \text{ on } D \end{aligned} \tag{4.3.1}$$

and the total wave function becomes

$$\psi(\mathbf{r})=\psi_i(\mathbf{r})+\int_J \mathrm{d}^2\rho\,\mathbf{n}_\rho\boldsymbol{\Gamma}_i(\mathbf{r},\boldsymbol{\rho}) \qquad \mathbf{r}\in\bar{\Omega}' \tag{4.3.2}$$

where $\boldsymbol{\Gamma}_i$ is the Green vector (4.1.8) with $\psi_s(\boldsymbol{\rho})$ replaced by $\psi_i(\boldsymbol{\rho})$.

An alternative form of $\psi(\mathbf{r})$ is obtained from the original integral (4.1.7) by defining a new surface A such that together with D it forces a closed surface $\Sigma'=A+D$ which encloses all sources of the wave field, i.e. those in Ω and the one at infinity guaranteeing $\psi_i(\mathbf{r})$ (Fig. 4.1). Then we can write instead of eqns (4.1.2) and (4.1.7)

$$\psi(\mathbf{r})=\int_{\Sigma^1} \mathrm{d}^2\rho\,\mathbf{n}_\rho\boldsymbol{\Gamma}(\mathbf{r},\boldsymbol{\rho}), \tag{4.3.3}$$

and by using assumptions (4.3.1) we obtain

$$\psi(\mathbf{r})=\int_A \mathrm{d}^2\rho\,\mathbf{n}_\rho\boldsymbol{\Gamma}_i(r,\rho) \qquad r\in\Omega'. \tag{4.3.4}$$

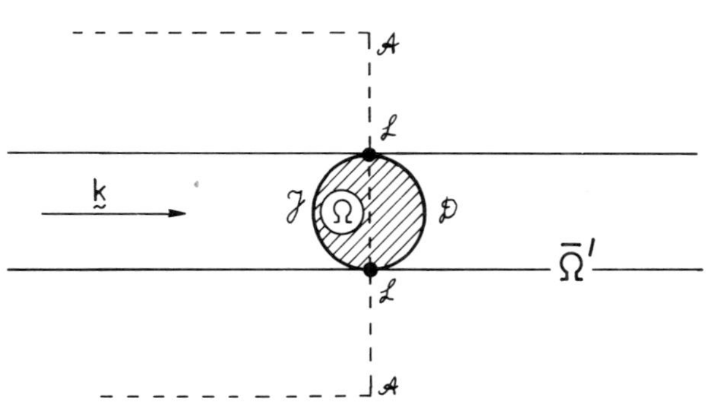

FIG. 4.1. Domains and surfaces in Kirchhoff's theory.

4.4 Reduction to a planar problem: Babinet's principle

Because the Green vector Γ_i satisfies the equation

$$\nabla_\rho \boldsymbol{\Gamma}_i(\mathbf{r}, \boldsymbol{\rho}) = -\psi_i(\boldsymbol{\rho})\,\delta(\mathbf{r}, \boldsymbol{\rho}) \tag{4.4.1}$$

it has zero divergence on the surfaces Σ or Σ'. Therefore the integrals over J in eqn (4.3.2) or over A in eqn (4.3.4) can be converted into line integrals along the 'shadow line' L separating J and D or A and D by virtue of Stokes' theorem. (In optics this conversion is known as the Maggi–Rubinowicz transformation.) As a result, in Kirchhoff's approximation (i.e. to the extent that the postulates (4.3.1) are valid), the field $\psi(\mathbf{r})$ depends only on the form of the shadow line. Thus the three-dimensional region Ω representing the 'black' scattering object can be replaced by an infinitely thin but still perfectly absorbing disc $\bar{A}$ bounded by L. The three-dimensional scattering problem is thereby reduced to a planar problem, and we define the 'diffraction plane' as the infinite plane that contains the shadow line L. Then the equivalence of eqns (4.3.2) and (4.3.4) means that the fields corresponding to two complementary areas A and $\bar{A} = D$ are related by

$$\psi^{(A)}(\mathbf{r}) + \psi^{(\bar{A})}(\mathbf{r}) = \psi_i(\mathbf{r}). \tag{4.4.2}$$

This relation is known as *Babinet's principle*. Although we have derived it here within Kirchhoff's approximation, it holds in general as a direct consequence of Huygens' principle.

4.5 Fresnel and Fraunhofer diffraction

Babinet's principle allows us to replace the problem of diffraction by a perfectly absorbing disc $\bar{A}$ by that of diffraction by the complementary aperture A in an infinitely extended perfectly absorbing plane screen (Fig. 4.2), and to calculate the field $\psi^{(A)}(\mathbf{r})$ from eqn (4.4.2) with $\bar{A} = A$. This yields, with $\mathbf{r}' = \mathbf{r} - \boldsymbol{\rho}$,

$$\psi^{(A)}(\mathbf{r}) = \frac{-ik}{4\pi} \int_A d^2\rho \{1 + \cos(\mathbf{n}_\rho, \mathbf{r}')\} \frac{\exp(ikr')}{r'}. \tag{4.5.1}$$

If we are interested only in the field at distances $r \gg a$ where a is the linear size of the aperture, we can replace $\mathbf{r}'$ by r in the slowly varying quantities in the integrand and write

$$\psi^{(A)}(\mathbf{r}) = \frac{-ik}{4\pi r}(1 + \cos\theta) \int_A d^2\rho \exp(ikr') \tag{4.5.2}$$

where θ is the angle between $\mathbf{n}_\rho$ and $\mathbf{r}$. Now we choose our coordinate

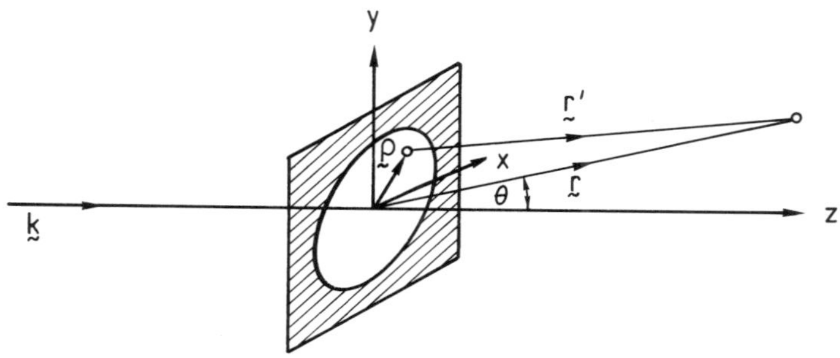

FIG. 4.2. Diffraction by an aperture.

systems such that the direction of $\mathbf{k}$ is the z axis and the diffraction plane is the plane $z=0$. By expanding r' about r to second order we obtain, with $\rho=(\xi, \eta, 0)$,

$$\psi^{(A)}(\mathbf{r})=\frac{-\mathrm{i}k}{4\pi}(1+\cos\theta)\frac{\mathrm{e}^{\mathrm{i}kr}}{r}\iint_A \mathrm{d}\xi\,\mathrm{d}\eta\,\exp[\mathrm{i}[\Phi(\xi,\eta)-(k'_x\xi+k'\eta)\}],$$

(4.5.3)

where

$$\Phi(\xi,\eta)=\frac{k}{2r}(\xi^2+\eta^2)-\frac{1}{2kr}(k'_x\xi+k'_y\eta)^2 \tag{4.5.4}$$

and where we have defined the vector ($\mathbf{k}'=k\hat{\mathbf{r}}$ of magnitude k in the direction of the observation point $\mathbf{r}$.

Now we can state the *conditions for diffraction in configuration space.* A general requirement for the validity of the Kirchhoff approximation is that the wavelength is small compared with the size of the diffracting object, i.e. $\lambda \ll a$ or

$$ka \gg 1. \tag{4.5.5}$$

In addition we distinguish between two different types of diffraction according to whether the quadratic terms $\Phi(\xi, \eta)$ in the phase of the integrand of (4.5.3) are negligible or significant in comparison with the linear terms; these are known as *Fraunhofer* and *Fresnel diffraction* respectively. In terms of the parameter

$$\rho = ka^2/r \tag{4.5.6}$$

eqns (4.5.3) and (4.5.4) yield the conditions

$$\begin{aligned}&\text{Fraunhofer diffraction:} && ka \gg 1 \quad \text{and} \quad \rho \ll 1\\ &\text{Fresnel diffraction:} && ka \gg 1 \quad \text{and} \quad \rho \gtrsim 1.\end{aligned} \tag{4.5.7}$$

Alternatively we can distinguish between the Fresnel and Fraunhofer case by the existence or absence respectively of a *point of stationary phase* (ξ_s, η_s) in the integrand of eqn (4.5.3). From eqns (4.5.2) and (4.5.3) we find $(\xi_s, \eta_s) = (x, y)/(\cos\theta)^2$, but the exact stationary point obtained from eqn (4.5.2) is $(\xi_s, \eta_s) = (x, y)$, the projection of the observation point onto the diffraction plane. The condition for Fresnel diffraction then becomes

$$\rho_s = r \sin\theta \lesssim a \quad \text{or} \quad \rho \gtrsim ka \sin\theta \lesssim 1 \tag{4.5.8}$$

while the Fraunhofer condition $\rho_s \gg a$ yields $\rho \ll ka \sin\theta \lesssim 1$ in agreement with (4.5.7).

4.6 Fraunhofer diffraction

In the Fraunhofer case we can neglect $\Phi(\xi, \eta)$, and we obtain

$$\psi^{(A)}(\mathbf{r}) = \frac{-ik}{4\pi}(1+\cos\theta)\frac{e^{ikr}}{r}\iint_A d\xi\, d\eta\, \exp\{-i(k'_x\xi + k'_y\eta)\}. \tag{4.6.1}$$

If we now go back to diffraction from the disc $\bar{A}$ by means of Babinet's principle (4.4.2), we can write

$$\psi^{(\bar{A})}(r) = \exp(i\mathbf{k}\cdot\mathbf{r}) + f(\mathbf{k}, \mathbf{k}')\frac{\exp(ikr)}{r} \tag{4.6.2}$$

with a *scattering amplitude* given by

$$f(\mathbf{k}, \mathbf{k}') = \frac{ik}{4\pi}(1+\cos\theta)\int_{\bar{A}} d^2\rho\, \exp(-i\mathbf{k}\cdot\boldsymbol{\rho}). \tag{4.6.3}$$

This can be generalized by writing

$$f(\mathbf{k}, \mathbf{k}') = \frac{ik}{4\pi}(1+\cos\theta)\int d^2\rho A(\boldsymbol{\rho})\exp(-i\mathbf{k}'\cdot\boldsymbol{\rho}) \tag{4.6.4}$$

where the integration extends over the whole diffraction plane, and the function $A(\boldsymbol{\rho})$ describes a more general source distribution than the 'sharp-cut-off' function

$$A_{sco}(\boldsymbol{\rho}) = \begin{cases} 1, & \boldsymbol{\rho} \in \bar{A} \\ 0, & \boldsymbol{\rho} \notin \bar{A} \end{cases} \tag{4.6.5}$$

which represents a sharp-edged perfectly absorbing disc. At this stage we also note that it is consistent with the general condition for the Kirchhoff approximation to replace the so-called 'obliquity factor' $1+\cos\theta$ by 2, since $\sin\theta \lesssim (ka)^{-1} \ll 1$. Then the scattering amplitude in the (Fraunhofer)

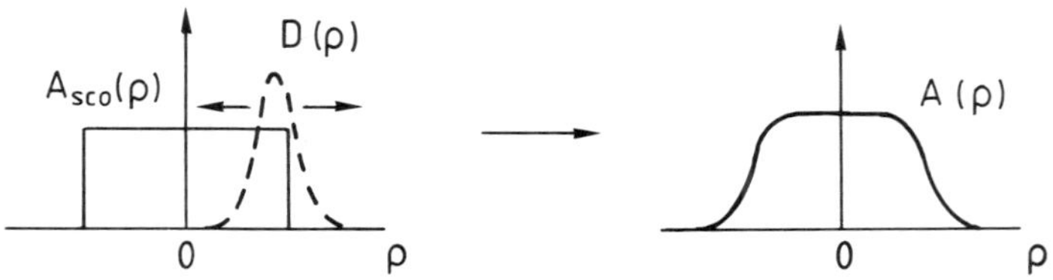

FIG. 4.3. Smoothing by convolution.

diffraction approximation (4.6.4) is, apart from a constant factor, the two-dimensional *Fourier transform of the distribution* function $A(\boldsymbol{\rho})$. If $A(\boldsymbol{\rho})$ is real, it represents an 'amplitude object', but if the scatterer is not perfectly absorbing and allows partial transmission of the wave imparting a phase shift to A, $A(\boldsymbol{\rho})$ will be complex and will represent a 'phase object' as well as an amplitude object. Such modifications are easily formulated in the diffraction representation. For instance, we can describe a smooth-edged scatterer by a distribution $A(\boldsymbol{\rho})$ obtained by convolution of the sharp-edged profile (4.6.5) with a smoothing function $D(\boldsymbol{\rho})$ as sketched in Fig. 4.3.

$$A(\boldsymbol{\rho}) = \int \mathrm{d}^2\rho' A_{\mathrm{sco}}(\boldsymbol{\rho}-\boldsymbol{\rho}')D(\boldsymbol{\rho}') \tag{4.6.6}$$

where D is normalized as $\int \mathrm{d}^2\rho D(\boldsymbol{\rho}) = 1$. Thus by a well-known property of the Fourier transformation the scattering amplitude can be written as the product

$$f(\mathbf{k}, \mathbf{k}') = f_{\mathrm{sco}}(\mathbf{k}, \mathbf{k}')F(\theta) \tag{4.6.7}$$

of the sharp-cut-off amplitude and the Fourier transform of the smoothing function

$$F(\theta) = \int \mathrm{d}^2\rho D(\boldsymbol{\rho})\exp(\mathrm{i}\mathbf{k}' \cdot \boldsymbol{\rho}). \tag{4.6.8}$$

If $D(\boldsymbol{\rho})$ is of the form shown in Fig. 4.3, $F(0)$ will smoothly decrease with increasing θ from its maximum $F(0) = 1$ at $\theta = 0$. Because it suppresses the amplitude of the sharp-cut-off oscillation towards large angles, $F(\theta)$ is called the 'damping factor'. The narrower is $D(\rho)$, the broader is $F(\theta)$; for example, a Gaussian

$$D(\rho) = (\pi d^2)^{-1/2} \exp(-\rho^2/d^2)$$

of width d yields a damping factor of width $1/d$.

If $A(\boldsymbol{\rho})$ has axial symmetry in the diffraction plane, the scattering amplitude (4.6.4) becomes, using polar coordinates (ρ, ϕ),

$$f(\theta)=\frac{ik}{2\pi}\int_0^\infty d\rho\,\rho A(\rho)\int_0^{2\pi} d\phi\,\exp(-ik\rho\sin\theta\cos\phi)$$
$$=ik\int_0^\infty d\rho\,\rho A(\rho)J_0(k\rho\sin\theta), \qquad (4.6.9)$$

where the integration over the azimuth ϕ has been carried out by means of an integral representation of the zero-order Bessel function

$$J_0(z)=\frac{1}{2\pi}\int_0^{2\pi} d\phi\,\exp(-iz\cos\phi). \qquad (4.6.10)$$

In this case $f(\theta)$ is related to $ikA(\rho)$ by what is known as a *Fourier–Bessel, or Hankel, transform* (in the variable $k\sin\theta$). If the scatterer is a spherical (or cylindrical) object represented by a sharp-edged disc of radius R, which may be partially transparent and impact a constant phase shift to the transmitted wave so that $A(\rho)=1-\alpha$ for $\rho\in\bar{A}$ with a complex transparency coefficient α, we can use the relation

$$\int dz\,zJ_0(z)=zJ_1(z) \qquad (4.6.11)$$

to obtain

$$f(\theta)=ik(1-\alpha)\int_0^R d\rho\,\rho J_0(k\rho\sin\theta)=i(1-\alpha)R\frac{J_1(kR\sin\theta)}{\sin\theta}. \qquad (4.6.12)$$

The corresponding intensity distribution ('Airy pattern')

$$I(z)=I_0\left\{\frac{2J_1(z)}{z}\right\}^2, \quad I_0=|f(0)|^2=|1-\alpha|^2\tfrac{1}{4}k^2R^4, \quad z=kR\sin\theta \qquad (4.6.12a)$$

is shown in Fig. 4.4.

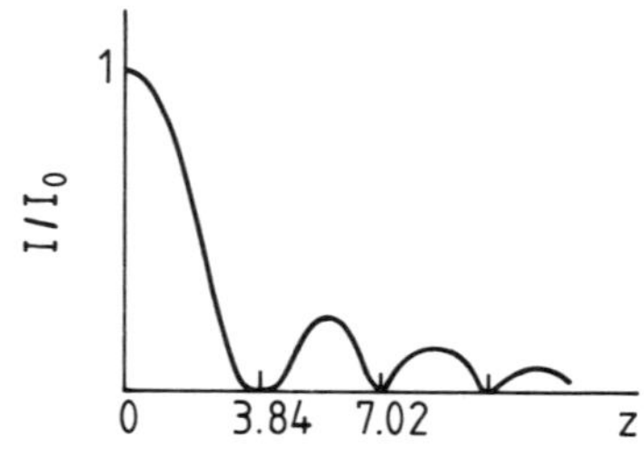

FIG. 4.4. Intensity distribution for Fraunhofer diffraction by a circular disc (Airy patterns). For large z the distance between diffraction minima is constant and equal to π.

4.7 Diffraction by a half-plane

The conditions for Fresnel diffraction in ordinary space imply that the observation point or the source of particles or both are at a *finite* distance from the scatterer. Since in subatomic collisions both the real source (e.g. the collimating slit of a particle accelerator) and the points of observation (detector) are always effectively at infinity compared with the size of the scattering region, it may seem that nuclear diffraction is always of the Fraunhofer type and that Fresnel diffraction can never occur. As we shall see in the theory of l-space diffraction (Chapter 5), this is not so: a strong enough *Coulomb* field can act, under certain conditions, like a diverging lens if the force is repulsive (it acts like a converging lens if the force is attractive) surrounding the region of nuclear interaction; as a result the incoming plane wave front is distorted in such a way that the particles appear to originate from a *virtual* source at a finite distance in front of the scatterer. Under these conditions Fresnel type diffraction occurs in much the same way as for a real source at finite distance, and it is instructive to compare the angular distribution derived from l-space diffraction with the intensity pattern obtained from the simplest case of Fresnel diffraction in ordinary space, i.e. scattering by a half-plane.

Because we have previously assumed an incoming plane wave (source at infinity), let us first consider the diffracted field of a perfectly absorbing half-plane of infinite extension in the y direction observed at finite distance $\mathbf{r}$ behind the sources (Fig. 4.5). Then we can confine ourselves to diffraction in the (x, z) plane. Going back to eqn (4.5.3) and setting $k'_y = 0$ so that $k_x'^2 + k_z'^2 = k^2$ and

$$\Phi(\xi, y) = \frac{k_z'^2}{2kr}\xi^2 + \frac{k}{2r}y^2, \tag{4.7.1}$$

we have

$$\psi^{(A)}(\mathbf{r}) = \frac{-ik\exp(ikr)}{2\pi\, r}\int_{-\infty}^{\infty} dy \exp\left(i\frac{k}{2r}y^2\right)\int_{-\infty}^{0} d\xi \times\exp\left\{i\left(\frac{k_z'^2}{2kr}\xi^2 - k'\xi\right)\right\}. \tag{4.7.2}$$

The integrations are easily performed with the result

$$\psi^{(A)}(\mathbf{r}) = \frac{\exp(ikr)}{\cos\theta}\exp\{-i\tfrac{1}{2}kr(\tan\theta)^2\}\tfrac{1}{2}\operatorname{erfc}\left\{\exp(-\tfrac{1}{4}\pi)\left(\frac{kr}{2}\right)^{1/2}\tan\theta\right\}, \tag{4.7.3}$$

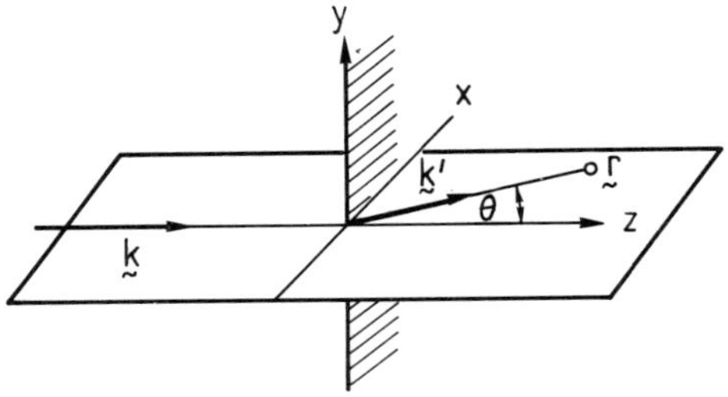

FIG. 4.5. Fresnel diffraction by a half-plane with observation point **r** at finite distance and source at infinity.

where the complementary error function is defined by

$$\operatorname{erfc}(z) = \frac{z}{\pi^{1/2}} \int_z^\infty \exp(-t^2)\,\mathrm{d}t. \tag{4.7.4}$$

However, in view of what has been said at the beginning of this section, it is more relevant for our purposes to consider the case of a point source at a finite distance from the half-plane and observation at infinity. This requires a generalization for which we go back to eqn (4.3.2) and let $\psi_i(\mathbf{r})$ represent a spherical wave originating from a source point $\mathbf{r}_0$:

$$\psi_i(\mathbf{r}) = \frac{1}{4\pi} \frac{\exp(ik\,|r - r_0|)}{|r - r_0|}. \tag{4.7.5}$$

Then the Kirchhoff formula (4.5.2) is generalized to

$$\psi^{(A)}(\mathbf{r}) = \frac{-ik}{4\pi} \int_A \mathrm{d}^2\rho \{\cos(\mathbf{n}_\rho, \mathbf{r}') - \cos(\mathbf{n}_\rho, \mathbf{r}_0')\} \frac{\exp\{ik(r_0' + r')\}}{r_0' r'} \tag{4.7.6}$$

For $r \gg a$ and $r' \gg a$ we can again approximate the slowly varying quantities in the integrand, $r_0' r' \approx r_0 r$ and $\cos(\mathbf{n}_\rho, \mathbf{r}') - \cos(\mathbf{n}_\rho, \mathbf{r}_0) \approx 2\cos(\mathbf{n}_\rho, \mathbf{r} - \mathbf{r}_0) 2 \cos\delta$, so that

$$\psi^{(A)}(r) = \frac{-ik\cos\delta}{2\pi\; r_0 r} \int_A \mathrm{d}^2\rho \exp\{ik(r_0' + r')\}. \tag{4.7.7}$$

By expanding r' about r and r_0' about r_0 to second order in the integrand, defining the vectors $\mathbf{k} = k\hat{\mathbf{r}}$ and $\mathbf{k}^0 = -k\hat{\mathbf{r}}_0$, and with the same choice of coordinate system as for eqn (4.5.3) we obtain

$$\psi^{(A)}(\mathbf{r}) = \frac{-ik}{2\pi} \cos\delta \frac{\exp\{ik(r_0 + r)\}}{r_0 r} \int \mathrm{d}\xi\, \mathrm{d}y \times \exp[i\{\Phi(\xi, y) - (k_x' - k_x^0)\xi - (k_y' - k_y^0)y\}] \tag{4.7.8}$$

where

$$\Phi(\xi, y)=\tfrac{1}{2}k\left(\frac{1}{r_0}+\frac{1}{r}\right)(\xi^2+y^2)-\frac{(k_x^0\xi+k_y^0y)^2}{2kr_0}-\frac{(k_x'\xi+k_y'y)^2}{2kr}. \qquad (4.7.9)$$

The conditions for Fresnel diffraction are now

$$\gamma \gtrsim 1 \quad \text{and/or} \quad \gamma_0 \gtrsim 1 \qquad (4.7.10)$$

where $\gamma_0 \equiv ka^2/r_0$. If we now let the observation point recede to infinity ($r \to \infty$) eqn (4.7.8) shows that we can then again define, at least formally, a *scattering amplitude*

$$f(\mathbf{k}_0, \mathbf{k}')=\frac{ik}{2\pi}\cos\delta\frac{\exp(ikr_0)}{r_0}\,d\xi\,dy$$
$$\times\exp[i\{\Phi_0(\xi, y)-(k_x'-k_x^0)\xi-(k_y'-k_y^0)y\}] \qquad (4.7.11)$$

with

$$\Phi_0(\xi, y)=\frac{k}{2r_0}(\xi^2+y^2)-\frac{(k_x^0\xi+k_y^0y)^2}{2kr_0}. \qquad (4.7.12)$$

For a half-plane in the y direction, with $k_y^0=k_y'=0$ and approximating the obliquity factor $\cos\delta \approx 1$, this reduces to

$$f(\mathbf{k}_0, \mathbf{k}')=\frac{ik}{2\pi}\frac{\exp(ikr_0)}{r_0}\int_{-\infty}^{\infty}d\eta\,\exp\left(i\frac{k}{2r_0}\eta^2\right)d\xi$$
$$\times\exp\left[i\left\{\frac{k_z^0}{2kr_0}\xi^2-(k_x'-k_x^0)\xi\right\}\right]. \qquad (4.7.13)$$

Carrying out the integrations in a similar way as for eqn (4.7.2) yields

$$f(\mathbf{k}_0, \mathbf{k}')=\frac{\exp(ikr_0)}{\cos\theta_0}\exp\left\{-i\tfrac{1}{2}kr_0\left(\frac{\sin\theta-\sin\theta_0}{\cos\theta_0}\right)^2\right\}$$
$$\times\tfrac{1}{2}\,\mathrm{erfc}\left\{\exp(-i\tfrac{1}{4}\pi)\left(\frac{kr_0}{2}\right)^{1/2}\frac{\sin\theta-\sin\theta_0}{\cos\theta_0}\right\}. \qquad (4.7.14)$$

For observation in the forward direction $\theta=0$ we have

$$f(\mathbf{k}_0, \mathbf{k}')=\frac{\exp(ikr_0)}{\cos\theta_0}\exp\{-i\tfrac{1}{2}kr_0(\tan\theta_0)^2\}$$
$$\times\tfrac{1}{2}\,\mathrm{erfc}\left\{-\exp(-i\tfrac{1}{4}\pi)\left(\frac{kr_0}{2}\right)^{1/2}\tan\theta_0\right\}. \qquad (4.7.15)$$

Comparison with eqn (4.7.3) shows the symmetry of the two diffraction problems with respect to interchanging source point and observation

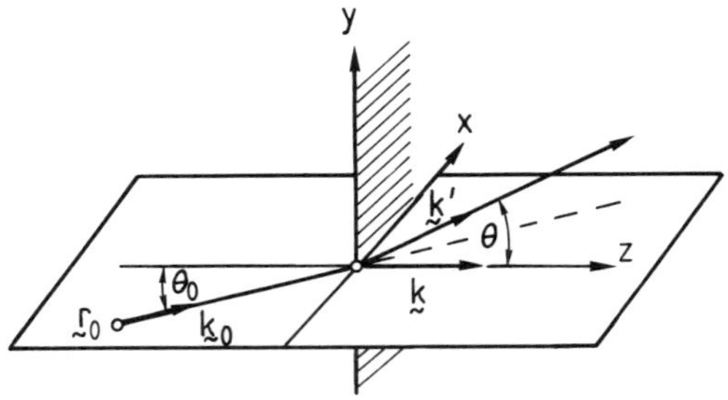

FIG. 4.6. Fresnel diffraction by a half-plane with the source point $\mathbf{r}_0$ at finite distance and observation at infinity.

point ($\mathbf{r}_0 \leftrightarrow \mathbf{r}$), a symmetry which is already incorporated in the general expressions (4.7.8) and (4.7.9) and is known as *Helmholtz's reciprocity theorem*. The only difference is in the sign of the argument of erfc(z), which expresses the fact that the 'illuminated' region for θ corresponds to the 'shadow' region for θ_0, and vice versa (Fig. 4.6).

However, we are more interested in the diffraction pattern near the boundary $\theta = \theta_0$ between the geometric-optical shadow and illuminated regions. For small $\theta - \theta_0$ we can approximate $(\sin\theta - \sin\theta_0)/\cos\theta_0 \approx 2\sin\frac{1}{2}(\theta-\theta_0)$, and then eqn (4.7.14) becomes

$$f(\mathbf{k}_0, \mathbf{k}') = \frac{\exp(\mathrm{i}kr_0)}{\cos\theta_0}\exp(-\mathrm{i}u^2)\tfrac{1}{2}\,\mathrm{erfc}\{\exp(-\mathrm{i}\tfrac{1}{4}\pi)u\} \qquad (4.7.16)$$

with

$$u = \left(\frac{kr_0}{2}\right)^{1/2} 2\sin\tfrac{1}{2}(\theta-\theta_0). \qquad (4.7.18)$$

Whenever the function erfc(z) appears with $\arg z = \pm\frac{1}{4}\pi$ or $\pm\frac{3}{4}\pi$, it signifies the presence of a Fresnel diffraction pattern. This is traditionally better known in terms of the Fresnel integrals

$$C(W) = \int_0^W \cos(\tfrac{1}{2}\pi t^2)\,\mathrm{d}t, \qquad S(W) = \int_0^W \sin(\tfrac{1}{4}\pi t^2)\,\mathrm{d}t \qquad (4.7.19)$$

to which erfc(z) is related by

$$\mathrm{erfc}\{\exp(\mathrm{i}\tfrac{1}{4}\pi)u\} = 1 - 2^{1/2}\exp(\mathrm{i}\tfrac{1}{4}\pi)\{C(W) - \mathrm{i}S(W)\} \qquad (4.7.20a)$$

$$\mathrm{erfc}\{\exp(-\mathrm{i}\tfrac{1}{4}\pi)u\} = 1 - 2^{1/2}\exp(\mathrm{i}\tfrac{1}{4}\pi)\{S(W) - \mathrm{i}C(W)\}, \qquad (4.7.20b)$$

with $W = (2/\pi)^{1/2}u$. For representing the intensity pattern $I = |f|^2$ it is convenient to define the 'Fresnel function'

$$\mathrm{Fr}(u) = |\tfrac{1}{2}\,\mathrm{erfc}\{\exp(\pm\mathrm{i}\tfrac{1}{4}\pi)u\}|^2 = \tfrac{1}{2}[\{\tfrac{1}{2} - C(W)\}^2 + \{\tfrac{1}{2} - S(W)\}^2], \qquad (4.7.21)$$

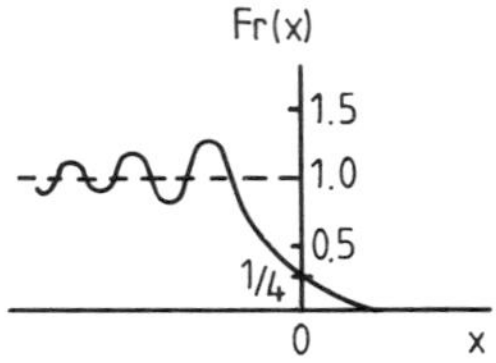

FIG. 4.7. The Fresnel function Fr(u).

whereby we can write

$$I(u)=I_0\,\mathrm{Fr}(u) \tag{4.7.22}$$

where $I_0=I(-\infty)$. The function Fr(u) is portrayed in Fig. 4.7, and shows the typical oscillations about unity in the 'illuminated region' ($u<0$) and the smooth decrease to zero in the 'shadow region' ($u>0$). Note that at the boundary $u=0$ between these regions $I(0)$ is *a quarter* of the full intensity I_0.

If the half-plane $0\leqslant x<\infty$ chosen in Figs. 4.5 and 4.6 is replaced by the complementary half-plane $-\infty<x\leqslant 0$, the arguments of the error function in eqns (4.7.3) and (4.7.16) change sign, interchanging illuminated and shadow regions. Babinet's principle is in this case simply expressed by a symmetry property of erfc(z):

$$\tfrac{1}{2}\,\mathrm{erfc}(z)+\tfrac{1}{2}\,\mathrm{erfc}(-z)=1. \tag{4.7.23}$$

Before considering the question of how to specify the functions $\sigma(\boldsymbol{\rho})$ and $t(\boldsymbol{\rho})$ of (4.2.1) in order to simulate nuclear interactions, we should comment on the basic assumption that there exists a region $\bar{\Omega}$ in which the interaction vanishes exactly. This already indicates the limitations of applying configuration-space diffraction theory to subatomic processes. Although the nuclear force is of short range and therefore nuclei have fairly well-defined surfaces, the nuclear potential never vanishes exactly. It extends to infinity in every direction and the 'tail' of the potential has important effects in nuclear collisions. Moreover, because the Coulomb interaction has infinite range, configuration-space diffraction theory strictly cannot be applied to charged particle collisions at all. Nevertheless, it may be expected to be a useful approximation at high energies when the wavelengths are small and the Coulomb interaction is relatively weak (we shall keep these restrictions in mind in making further approximations).

4.8 Reference

[4.1] Hönl, H., Mane, A. W., and Westpfahl, K. (1961). *Handbuch der Physik* **25,** 1 (Springer, Berlin).

5

DIFFRACTION IN ANGULAR MOMENTUM SPACE

5.1 Introduction

By 'quantum diffraction in angular momentum space', we refer to that diffraction associated with the uncertainty relation between orbital angular momentum components and conjugate angles. Because the spectrum of $\mathbf{L}_z$ is discrete and bounded, the commutator $[\phi, L_z] = i\hbar$ and the corresponding uncertainty relation $\Delta\phi\Delta L_z \geqslant \hbar/2$ are not strictly valid; correct forms can be given in which the azimuthal angle variable ϕ is replaced by the continuous periodic variables $\sin\phi$ and $\cos\phi$. However, in the near-classical regime of large angular momenta, the spectra of $\mathbf{L}_z$ and L^2 can be regarded as quasi-continuous and the simpler form of the uncertainty relation holds to a good approximation. Moreover, for $L \gg \hbar$ there exists a reciprocity between the magnitude of the orbital angular momentum and the polar angle θ, as shown by the asymptotic form of the Legendre polynomials $P_l(\cos\theta) \simeq J_0\{(l+\frac{1}{2})\theta\}$, which can be expressed by an approximate uncertainty relation $\Delta\theta\Delta L \geqslant \hbar$, where $L \approx (l+\frac{1}{2})\hbar$.

This is the case at high energies, and many collision processes in nuclear and particle physics at high enough energy can be regarded as bandpass filtration measurements in angular momentum space whenever the scatterer is strongly absorbing and spatially extended with a well-defined surface. The corresponding intensity distributions as a function of scattering angle (differential cross-sections) show a pronounced diffraction structure, particularly in nuclear scattering and reactions. This gave rise to the development of phenomenological diffraction models for nuclear collisions.

5.2 Elastic scattering

The point of departure is the expansion of the scattering amplitude in eigenstates of angular momentum, i.e. for a spherically symmetrical interaction the partial-wave expansion

$$f(\theta) = \frac{i}{k} \sum_{l=0}^{\infty} (l+\tfrac{1}{2})(1-S_l)P_l(\cos\theta). \qquad (5.2.1)$$

If the characteristic angular momentum $\bar{L}=\hbar(l+\frac{1}{2})=pR$ corresponding to the radius R of the interaction region is large ($L_c \gg \hbar$) the summation over l can be replaced by the integral

$$f(\theta)=\hbar^{-1}\int_0^\infty g(L)J_0(L\theta/\hbar)L\,\mathrm{d}L \tag{5.2.2}$$

which represents the scattering (θ-space) amplitude as a Fourier–Bessel transform of the partial wave (L-space) amplitude

$$g(L)=\frac{\mathrm{i}}{p}\left\{1-S\left(\frac{L}{\hbar}\right)\right\}.$$

Equation (5.2.2) corresponds to Fraunhofer diffraction in L-space by a spherical scatterer; in the simplest case where

$$g_\mathrm{D}(L)=\frac{\mathrm{i}}{p}\,\mathrm{circ}\left(\frac{L}{\bar{L}}\right)$$

the intensity distribution (i.e. the differential scattering cross-section) $\sigma_\mathrm{D}(\theta)=|f_\mathrm{D}(\theta)|^2$ is an Airy pattern

$$\frac{\sigma_\mathrm{D}(\theta)}{\sigma_\mathrm{D}(0)}=\left\{\frac{2J_1(\bar{L}\theta/\hbar)}{\bar{L}\theta/\hbar}\right\}^2. \tag{5.2.3}$$

Fraunhofer diffraction occurs in the scattering of high-energy neutrons or of light nuclei when Coulomb effects are negligible. In the presence of Coulomb interaction, eqn (5.2.1) changes to

$$f(\theta)=f_c(\theta)+\frac{\mathrm{i}\hbar}{2p}\sum_{l=0}^{\infty}(2l+1)(1-S_l)\exp(\mathrm{i}2\sigma_l)P_l(\cos\theta) \tag{5.2.4}$$

where $f_c(\theta)$ is the Rutherford scattering amplitude and the σ_l are the Coulomb phase shifts. Instead of eqn (5.2.2) we can write approximately

$$\varphi_\mathrm{D}(\theta)=(2\pi\hbar)^{-1/2}\int\psi_\mathrm{D}(L)\exp(-\mathrm{i}L\theta/\hbar)\,\mathrm{d}L \tag{5.2.5}$$

where $\varphi_\mathrm{D}(\theta)=f(\theta)-f_c(\theta)$ and

$$\psi_\mathrm{D}(L)=\left(\frac{L}{\sin\theta}\right)^{1/2}g_\mathrm{D}(L)\exp(\mathrm{i}S(L)/\hbar)$$
$$\frac{S(L)}{\hbar}=2\sigma(L/\hbar)+\pi/4. \tag{5.2.6}$$

For large values of the Sommerfeld parameter n the asymptotic expression of the Coulomb phases is

$$2\sigma(L/\hbar)=2\sigma_0-2n\ln(\sin\theta/2)+n\theta\cot\tfrac{1}{2}\theta-\pi/2 \tag{5.2.7}$$

where $\theta = 2\arctan(n\hbar/L)$, and the integrand in eqn (5.2.5) has a point of stationary phase at $L_0 = n\hbar \cot \frac{1}{2}\theta$.

Under the conditions $n \gg 1$, $\bar{L} \gg \hbar$ at $n\hbar/\bar{L} = \text{constant}$, evaluation of the integral (5.2.5) according to the general method of Sections 4.5 and 4.6 yields for the ratio of the differential cross-section $\sigma(\theta)$ to the Rutherford cross-section $\sigma_R(\theta) = |f_c(\theta)|^2$,

$$\frac{\sigma(\theta)}{\sigma_R(\theta)} = \frac{1}{4}\,|\text{erfc}\{A(\theta, \bar{\theta})\}|^2 \tag{5.2.8}$$

with

$$A(\theta, \bar{\theta}) = \exp\left(-\mathrm{i}\frac{\pi}{4}\right)\frac{\theta - \bar{\theta}}{\{2\hbar S''(\bar{L})\}^{1/2}} = \exp\left(\frac{\mathrm{i}\pi}{4}\right)\left(\frac{\bar{L}}{2\hbar \sin\bar{\theta}}\right)^{1/2}(\theta - \bar{\theta}) \tag{5.2.9}$$

where $\bar{\theta} = S'(\bar{L}) = 2\arctan(n\hbar/\bar{L})$ and $S''(\bar{L}) = -(2/n\hbar)(\sin\bar{\theta}/2)^2 = -\sin\bar{\theta}/\bar{L}$.

Thus for a strong enough Coulomb interaction, nuclear diffractive scattering changes from the Fraunhofer to the Fresnel type. There are simple criteria for the occurrence of the two types of patterns analogous to the conditions given in Section 4.5:

$$\begin{array}{lll} \bar{L} \gg \hbar, & \bar{L}\sin\bar{\theta} \ll \hbar & \text{Fraunhofer diffraction} \\ \bar{L} \gg \hbar, & \bar{L}\sin\bar{\theta} \gtrsim \hbar & \text{Fresnel diffraction,} \end{array} \tag{5.2.10}$$

where $y_0 \equiv \bar{L}\sin\bar{\theta}$ is the analogue of the Rayleigh parameter. With these criteria one can give a general classification of all collision processes in nuclear and particle physics with respect to their diffraction characteristics. As we emphasize in Chapter 6, Fresnel-type scattering is predominant in heavy-ion collisions at small angles.

Closely related to the angular momentum representation of diffractive scattering at high energies is the impact parameter representation which forms a link between quantum diffraction in L-space and in coordinate space. The connection is given by the semiclassical relation $L \approx pb$, where b is the impact parameter. In this representation the scattering amplitude is given by

$$f_D(\mathbf{p} - \mathbf{p}') = \frac{ip}{2\pi\hbar}\int \Gamma_D(b)\exp\left[\frac{\mathrm{i}}{\hbar}(\mathbf{p} - \mathbf{p}')\cdot\mathbf{b}\right]\mathrm{d}^2 b \tag{5.2.11}$$

where the integral extends over two-dimensional impact parameter space, i.e. a plane perpendicular to the incoming momentum **P**. Thus the scattering amplitude is essentially the two-dimensional Fourier transform of the 'profile function':

$$\Gamma_D(\mathbf{b}) = 1 - S(\mathbf{b}) = 1 - \exp\{\mathrm{i}2\delta(\mathbf{b})\} \tag{5.2.12}$$

where $\delta(\mathbf{b})$ is the phase shift function which is a continuous approximation of the scattering phase shifts δ_l. Since the momentum transfer $\mathbf{q}=\mathbf{p}-\mathbf{p}'$ lies approximately in the transverse plane to $\mathbf{p}$, eqn (5.2.11) is essentially the same as eqn (4.6.9). For a circularly symmetrical profile function, eqn (5.2.11) becomes the Fourier–Bessel transform relation

$$f_{\mathrm{D}}(q)=\frac{\mathrm{i}}{\hbar}p\int_0^{\infty}\Gamma_{\mathrm{D}}(b)J_0\left(\frac{bq}{\hbar}\right)b\,\mathrm{d}b \qquad (5.2.13)$$

which is analogous to eqn (5.2.2).

In nuclear and hadronic collisions the profile function is determined by the structure and composition of the scattering target as well as of the projectile. For composite systems the profile function can be expanded in a multiple scattering series, and the scattering amplitude $f_{\mathrm{D}}(q)$ is represented in terms of the amplitudes for scattering of the individual constituents of the collision partners. Multiple diffraction theories of the Glauber type have been most successful in describing high-energy collisions of nucleons, pions, etc. by nuclei.

The purpose of this rather short chapter is to give a general framework for the discussion that follows in the next chapters.

PART II

CLOSED FORMALISM FOR STRONG-ABSORPTION PROCESSES

6

ELASTIC SCATTERING

6.1 Preview

In this part we endeavour to give a coherent account of a closed-form description of strong-absorption processes. This description is based on general properties of the partial-wave S-matrix S_l under conditions of strong absorption, and provides techniques for evaluating partial-wave summations in the scattering and reaction amplitudes in explicit analytic form. The result is a consistent formulation of quantum diffraction (see Section 1.2) in orbital angular momentum space, in which restrictions on the spectrum of l-values give rise to diffraction patterns in the conjugate angular distributions.

Chapter 6 deals with elastic scattering. This is followed by a description of inelastic scattering (Chapter 7) and transfer reactions (Chapter 8) of spin-zero particles. The extension to coupled channels is described separately in Chapter 9. These chapters form the central part of this book and are its main *raison d'être*.

6.2 Partial-wave S-matrix

Our starting point is the partial-wave representation of the elastic scattering amplitude for spin-zero particles:

$$f(\theta)=\frac{\mathrm{i}}{k}\sum_{l=0}^{\infty}(l+\tfrac{1}{2})(1-S_l)P_l(\cos\theta). \tag{6.2.1}$$

In the following developments we regard the S-matrix elements $S_l = \eta_l \exp(\mathrm{i}2\delta_l)$ as *given* quantities, generated by microscopic interaction models or by phenomenological optical potentials. We exploit the result of many numerical calculations that under conditions of strong absorption (and in the absence of specific dynamical effects giving rise to 'anomalous' scattering which will be discussed in Chapter 9) the reflection coefficients η_l and phase shifts δ_l have a simple smooth dependence on l as well as on energy E. A typical example of η_l and the 'quantal deflection function' $\Theta_l = 2(\delta_{l+1}-\delta_l) \equiv \mathrm{d}(2\delta_l)/\mathrm{d}l$ is shown in Fig. 6.1. The reflection coefficients for the low-l partial waves are practically zero up to a narrow transition

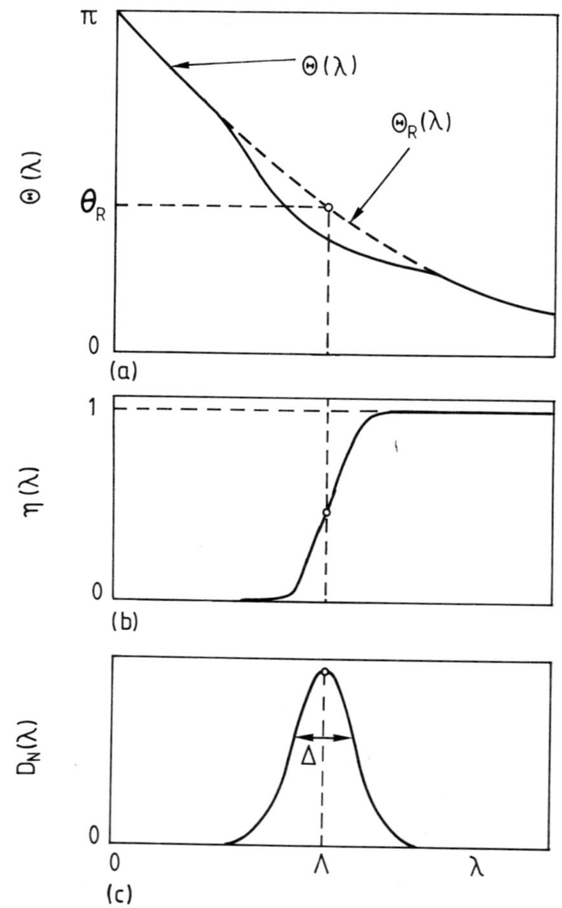

FIG. 6.1. (a) Quantal deflection function $\Theta(\lambda)$ (– – –, Rutherford deflection function $\Theta_R(\lambda)$), (b) reflection function $\eta(\lambda)$, and (c) absorptive shape function $D_N(\lambda)$ for strong-absorption scattering.

region around a critical angular momentum l_0 over which η_l rises smoothly to unity. Normally the deflection function Θ_l also varies smoothly with l, having its maximum deviation from the Rutherford values, $\Theta_{N,l} = \Theta_l - \Theta_{R,l}$, in the vicinity of l_0; in any case the variation of Θ_l at low l-values is irrelevant because of strong absorption. We characterize this generic l-dependence by saying that S_l has a 'normal strong-absorption profile'.

Under the general condition for diffraction scattering, $l_0 \gg 1$ many partial waves take part in the scattering process and the summation in eqn (6.2.1) involves only large l-values. Because of their smooth variation, the quantities η_l and δ_l can be regarded as (differentiable) functions of the continuous variable $\lambda = l + \frac{1}{2}$, and we shall replace S_l by the 'scattering (S-)function'

$$S(\lambda) = \eta(\lambda)\exp\{i2\delta(\lambda)\} = S_N(\lambda)\exp\{i2\sigma(\lambda)\}, \tag{6.2.2}$$

where

$$S_N(\lambda) = \eta(\lambda)\exp\{i2\delta_N(\lambda)\} \tag{6.2.3}$$

is the 'nuclear' part of the S-function, and is defined as the deviation from pure Rutherford (i.e. point-charge Coulomb) interaction. As we shall see, the nuclear interaction enters essentially through the *Fourier transforms of the derivative of* $S_N(\lambda)$. Because of its normal strong-absorption profile the magnitude of the function

$$D_N(\lambda) = \frac{dS_N(\lambda)}{d\lambda} = \{D(\lambda) + i\Theta_N(\lambda)\eta(\lambda)\}\exp\{i2\delta_N(\lambda)\}, \tag{6.2.4}$$

where $D(\lambda) = d\eta(\lambda)/d\lambda$ and $\Theta_N(\lambda) = d\{2\delta_N(\lambda)\}/d\lambda$, is confined to a narrow domain centred on $\Lambda = l_0 + \frac{1}{2}$. This domain can be regarded as the l-space 'window' through which the elastic scattering proceeds. If its width is measured by a suitably defined 'window parameter' Δ, we can assume that the functions $\eta(\lambda)$, $S_N(\lambda)$ and $D_N(\lambda)$ depend on λ, Λ and Δ in the combination $\mu = (\lambda - \Lambda)/\Delta$, so that we can write

$$\eta(\lambda) = \hat{\eta}(\mu), \qquad S_N(\lambda) = \hat{S}_N(\mu), \qquad D_N(\lambda) = \hat{D}_N(\mu). \tag{6.2.5}$$

The critical value $\Lambda = l_0 + \frac{1}{2}$, corresponding to the 'grazing angular momentum' l_0, has been defined in different ways, such as $\eta(\Lambda_{1/2}) = \frac{1}{2}$, or $\mathrm{Re}\, S_N(\Lambda_r) = \frac{1}{2}$, or in terms of the transmission coefficient as $T(\Lambda_T) = 1 - \{\eta(\Lambda_T)\}^2 = \frac{1}{2}$ so that $\eta(\Lambda_T) = 1/\sqrt{2}$. However, because of the significance of the function $D_N(\lambda)$ through the Fourier transforms, the most suitable definition of Λ is as the maximum of $|D_N(\lambda)|$ (Frahn 1978 [6.1]); we shall denote

$$|D_N(\lambda)| = [\{D(\lambda)\}^2 - \{\Theta_N(\lambda)\eta(\lambda)\}^2]^{1/2} \tag{6.2.6}$$

the 'absorptive shape function'. If this is a symmetrical function, our definition of Λ coincides with $\Lambda_{1/2}$, but normally it differs slightly from the latter value since in general $|D_N(\lambda)|$ has an asymmetric shape.

Although our formulation treats $S_N(\lambda)$ as a smooth function, we explicitly allow for discontinuous and discrete l-values as well. In fact, $S_N(\lambda)$ can be any generalized function (distribution) defined over any domain in real λ- or l-space and subject to correspondingly generalized operations of differentiation etc., the only requirement being that $S_N(\lambda)$ or its derivative have well-defined Fourier transforms. For this reason, a recent formulation of strong-absorption scattering (Rowley 1980 [6,2]), in which the discreteness of S_l is explicitly retained, gives no more general results than the formalism to be described here.

We wish to emphasize that aside from the general assumptions on $S_N(\lambda)$, i.e. its strong-absorption profile and dependence on $\mu=(\lambda-\Lambda)/\Delta$, its form need not be specified further. It appears in the expressions for the scattering amplitude only by way of Fourier transforms of $D_N(\lambda)$, whose properties can be discussed without specifying its functional form. This is a major advantage of the present formalism, which is based on the distribution of S_l over real l-space and allows for generalized functions, compared with Regge-type descriptions which require specific functional models of $S(\lambda)$ for analytic continuation into the complex λ-plane.

However, it is often instructive to illustrate the general results by choosing an analytic form of $S_N(\lambda)$. Convenient parametrizations are those of McIntyre *et al.* (1960 [6.3]) and Ericson (1965 [6.4]). Throughout the following chapters we adopt the latter, which is given by

$$\hat{S}_N(\mu)=\{1+\exp(-\mu-i\alpha)\}^{-1}, \qquad \mu=(\lambda-\Lambda)/\Delta. \tag{6.2.7}$$

The corresponding reflection function and phase shift function are

$$\hat{\eta}(\mu)=\{1+2(\cos\alpha)e^{-\mu}+e^{-2\mu}\}^{-1/2} \tag{6.2.8}$$

and

$$2\hat{\delta}_N(\mu)=\arctan\left(\frac{\sin\alpha}{\cos\alpha+e^{\mu}}\right); \tag{6.2.9}$$

hence the absorptive shape function and quantal deflection function become

$$\hat{D}_N(\mu)=\frac{1}{\Delta}\frac{\exp(-\mu-i\alpha)}{\{1+\exp(-\mu-i\alpha)\}^2} \tag{6.2.10}$$

and

$$\hat{\Theta}_N(\mu)=-\frac{1}{\Delta}\frac{(\sin\alpha)e^{-\mu}}{1+2(\cos\alpha)e^{-\mu}+e^{-2\mu}}. \tag{6.2.11}$$

Thus the nuclear phase shifts are mainly determined by the parameter α. The functions $\hat{\eta}(\mu)$ and $\hat{\delta}_N(\mu)$ are monotonic for $|\alpha|<\frac{1}{2}\pi$, and for attractive nuclear interaction we restrict α to the interval $0\leqslant\alpha\leqslant\frac{1}{2}\pi$. Note that for the form (6.2.7) the maximum of $|D_N(\lambda)|$ is at $\lambda=\Lambda$, in accordance with our general definition of this parameter.

Because of the generality of our assumptions on the S-matrix there is no compelling need to address the difficult question of how the properties of $S_N(\lambda)$ are determined by the interaction that generates this function, or, more specifically, how the l-space parameters Λ, Δ, and α are related to potential parameters in coordinate space, i.e. radii, diffusenesses, and the depths of the real and imaginary potential wells. In particular this question involves the continuous ambiguities in the potential parameters that are characteristic of strong-absorption interactions; such ambiguities are not present, or at least to a much lesser extent, for the l-space parameters. However, we do not entirely bypass this problem; approximate relations between S-matrix and interaction potentials are discussed in Appendix 1. Such relations are mainly useful for giving clues to the energy dependence of the S-matrix parameters, which is needed for a discussion of excitation functions.

6.3 'Neutral' particles

The main application of the formalism described in Chapters 6–9 is to the scattering and direct reactions of complex nuclei, particularly of heavy ions above the Coulomb barrier, whose interactions are characterized by both strong absorption and strong Coulomb fields. However, it is instructive first to consider situations in which the Coulomb interaction is negligibly weak, under conditions of the 'neutral' limit ($\mathcal{N}$-limit) defined in Section 5.2. Then the diffractive processes are purely of Fraunhofer type, and this applies to the scattering of light-ion projectiles: alpha particles, helions, tritons and deuterons, and (to some extent) nucleons at higher energies.

To begin with, in treating elastic scattering we shall always exclude the singularity at $\theta = 0$ in the partial-wave series (6.2.1) which arises from

$$\sum_{l=0}^{\infty} (l+\tfrac{1}{2})P_l(\cos\theta) = \delta(\cos\theta - 1), \tag{6.3.1}$$

so that we can write

$$f(\theta) = -\frac{\mathrm{i}}{k}\sum_{l=0}^{\infty} (l+\tfrac{1}{2})S_l P_l(\cos\theta). \tag{6.3.2}$$

The summation over l can be replaced by a sum of integrals over $\lambda = l + \frac{1}{2}$ by means of Poisson's summation formula:

$$f(\theta) = -\frac{\mathrm{i}}{k}\sum_{m=-\infty}^{\infty} \exp(-\mathrm{i}m\pi)\int_0^{\infty} \mathrm{d}\lambda\lambda S_N(\lambda)P_{\lambda-1/2}(\cos\theta)\exp(\mathrm{i}2m\pi\lambda), \tag{6.3.3}$$

where we have, for 'neutral' particles, neglected the Coulomb phase shifts in the S-function $S(\lambda)$. In a semiclassical picture the integer m describes the number of times the projectile's trajectory encircles the scattering centre. This already suggests that under strong-absorption conditions, and for all scattering angles not close to π, the term with $m=0$ will be dominant by far over all other contributions in the Poisson series. However, in this section we shall retain the full Poisson series in order to determine the relative magnitude of the $m=0$ and $m\neq 0$ terms.

In terms of the quantities Λ and Δ characterizing the profile of $S_{\mathrm{N}}(\lambda)$, our general conditions of diffraction and strong absorption can be stated as

$$\Lambda \gg 1, \qquad \Delta \ll \Lambda. \tag{6.3.4}$$

Therefore we can replace the Legendre polynomials in eqn (6.3.3) by their asymptotic forms for large l. Depending on our purpose we shall use one of the following expressions:

$$P_{\lambda-1/2}(\cos\theta) \simeq \left(\frac{\theta}{\sin\theta}\right)^{1/2} J_0(\lambda\theta), \qquad 0 \leqslant \theta \leqslant \pi - \lambda^{-1} \tag{6.3.5}$$

$$P_{\lambda-1/2}(\cos\theta) \simeq \exp\{\mathrm{i}\pi(\lambda-\tfrac{1}{2})\}\left(\frac{\pi-\theta}{\sin\theta}\right)^{1/2} J_0\{\lambda(\pi-\theta)\}, \qquad \lambda^{-1} \leqslant \theta \leqslant \pi \tag{6.3.6}$$

$$P_{\lambda-1/2}(\cos\theta) \simeq \left(\frac{2}{\pi\lambda\sin\theta}\right)^{1/2} \cos(\lambda\theta - \tfrac{1}{4}\pi), \qquad \lambda^{-1} \leqslant \theta \leqslant \pi - \lambda^{-1}. \tag{6.3.7}$$

The form (6.3.5) will be used for 'small-angle' scattering, down to and including $\theta=0$, and the form (6.3.6) for 'large-angle' scattering, up to and including $\theta=\pi$. When the regions close to $\theta=0$ and $\theta=\pi$ are not of particular interest we can use the third form (6.3.7).

With the small-angle asymptotics (6.3.5) we can write

$$f(\theta) = -\frac{\mathrm{i}}{k}\left(\frac{\theta}{\sin\theta}\right)^{1/2} \sum_{m=-\infty}^{\infty} \exp\{\mathrm{i}2m\pi(\Lambda-1/2)\} I_m(\theta) \tag{6.3.8}$$

where

$$I_m(\theta) = \int_0^\infty \mathrm{d}\lambda\,\lambda S_{\mathrm{N}}(\lambda) J_0(\lambda\theta) \exp\{\mathrm{i}2m\pi(\lambda-\Lambda)\}. \tag{6.3.9}$$

These integrals (in which we do not explicitly indicate a suitable convergence factor to ensure proper behaviour of the integrand at large λ) have the form of combined Fourier and Bessel transforms of $S_{\mathrm{N}}(\lambda)$, for which a general evaluation method is described in Appendix 2. Use of eqn. (II.3) gives the result

$$I_m(\theta)=\mathrm{i}\tfrac{1}{2}\Lambda\left[\frac{F_N\{\Delta(2m\pi-\theta)\}}{2m\pi-\theta}\{J_0(\Lambda\theta)-\mathrm{i}J_1(\Lambda\theta)\}\right.$$
$$\left.+\frac{F_N\{\Delta(2m\pi+\theta)\}}{2m\pi+\theta}\{J_0(\Lambda\theta)+\mathrm{i}J_1(\Lambda\theta)\}\right]$$

where the function $F_N(\Delta z)$ is defined as the Fourier transform of $D_N(\lambda)=\mathrm{d}S_N(\lambda)/\mathrm{d}\lambda$,

$$F_N(\Delta z)=\int_{-\infty}^{\infty}\mathrm{d}\lambda D_N(\lambda)\exp\{\mathrm{i}(\lambda-\Lambda)z\}=\Delta\int_{-\infty}^{\infty}\mathrm{d}\mu\hat{D}_N(\mu)\exp(\mathrm{i}\mu\Delta z). \tag{6.3.11}$$

Thus we can write

$$f(\theta)=\frac{\Lambda}{2k}\left(\frac{\theta}{\sin\theta}\right)^{1/2}\left[G_{SA}^{(0)}(\theta)\{J_0(\Lambda\theta)-\mathrm{i}J_1(\Lambda\theta)\}\right.$$
$$\left.+G_{SA}^{(0)}(-\theta)\{J_0(\Lambda\theta)+\mathrm{i}J_1(\Lambda\theta)\}\right], \tag{6.3.12}$$

where

$$G_{SA}^{(0)}(\pm\theta)=\sum_{m=-\infty}^{\infty}\exp\{\mathrm{i}2m\pi(\Lambda-\tfrac{1}{2})\}\frac{F_N\{\Delta(2m\pi\mp\theta)\}}{2m\pi\mp\theta}. \tag{6.3.13}$$

The subscript SA stands for small angle, but eqn (6.3.12) is valid for all angles except for a small region near $\theta=\pi$, subject to the general condition $\Delta^2\theta/\Lambda\ll 1$ which is always very well satisfied under strong-absorption conditions.

The form of eqn (6.3.12) is quite transparent. The amplitude has two branches, $f^{(+)}(\theta)$ and $f^{(-)}(\theta)$, which in a semiclassical picture correspond to scattering from the 'near side' and the 'far side' of the interaction region respectively. This can be seen from the asymptotic form of the Bessel function combinations,

$$J_0(\Lambda\theta)\mp\mathrm{i}J_1(\Lambda\theta)\simeq\left(\frac{2}{\pi\Lambda\theta}\right)^{1/2}\exp\{\mp\mathrm{i}(\Lambda\theta-\tfrac{1}{4}\pi)\}; \tag{6.3.14}$$

if the (+) branch scatters through θ, the (−) branch scatters through -2θ, as shown in Fig. 6.2. The relative magnitude of the two branches is determined by the functions $G_{SA}^{(0)}(\pm\theta)$ depending on the shape of $S_N(\lambda)$ or $D_N(\lambda)$.

An alternative form of our result is obtained by introducing the functions

$$\begin{aligned}\bar{f}_{SA}(\theta)&=\tfrac{1}{2}\theta\{G_{SA}^{(0)}(-\theta)-G_{SA}^{(0)}(\theta)\}\\ \tilde{f}_{SA}(\theta)&=\tfrac{1}{2}\{G_{SA}^{(0)}(-\theta)+G_{SA}^{(0)}(\theta)\},\end{aligned} \tag{6.3.15}$$

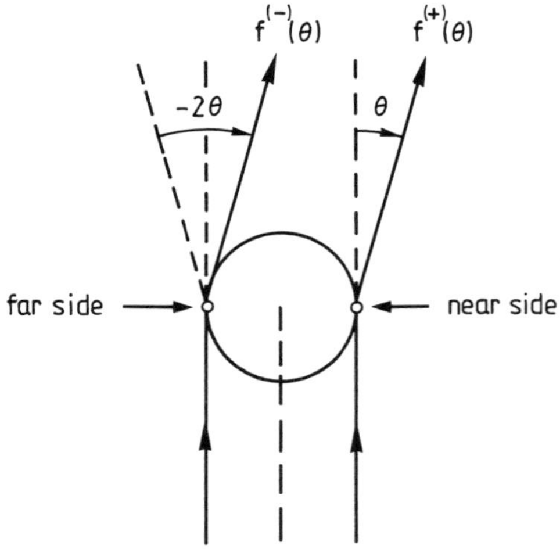

FIG. 6.2. 'Near-side' and 'far-side' scattering to forward angles.

with which eqn (6.3.12) can be written as

$$f(\theta)=\frac{\Lambda}{k}\left(\frac{\theta}{\sin\theta}\right)^{1/2}\left\{\mathrm{i}\bar{f}_{\mathrm{SA}}(\theta)\frac{J_1(\Lambda\theta)}{\theta}+\tilde{f}_{\mathrm{SA}}(\theta)J_0(\Lambda\theta)\right\}. \qquad (6.3.16)$$

The first term is the amplitude of a quantal Airy pattern, and is analogous to Fraunhofer diffraction by an obstacle of circular cross-section with radius $R=\Lambda/k$. The function $\bar{f}_{\mathrm{SA}}(\theta)$ results in a damping of the Airy pattern by the factor $|\bar{f}_{\mathrm{SA}}(\theta)|^2$; this damping becomes stronger with increasing values of Δ. The decond term, which is proportional to $J_0(\Lambda\theta)$, oscillates out of phase with the first term and causes a filling-in of the minima in the Airy distribution. We shall see shortly that this contribution is essentially determined by the nuclear phase shifts. It can therefore be regarded as the 'refractive' part of neutral particle scattering, in contrast with the 'diffractive' part described by the first term.

To illustrate these features we adopt the parametrization (6.2.7) for which the Fourier transform (6.3.11) has the simple form

$$F_{\mathrm{N}}(\Delta z)=F(\Delta z)\exp(\alpha\Delta z), \qquad F(\Delta z)=\frac{\pi\Delta z}{\sinh(\pi\Delta z)}. \qquad (6.3.17)$$

Then we have

$$G_{\mathrm{SA}}^{(0)}(\pm\theta)=\pi\Delta\exp(\mp\alpha\Delta\theta)\sum_{m=-\infty}^{\infty}\frac{\exp[\mathrm{i}\{2m\pi(\Lambda-\frac{1}{2}-\mathrm{i}\alpha\Delta)\}]}{\sinh\{\pi\Delta(2m\pi\mp\theta)\}}. \qquad (6.3.18)$$

After singling out the leading $m=0$ terms, the remaining $m\neq 0$ terms can be summed (for not too small Δ) with the result

$$G^{(0)}_{\mathrm{SA}}(\pm\theta)=\mp\pi\Delta\exp(\mp\alpha\Delta\theta)\times\left\{\frac{1}{\sinh(\pi\Delta\theta)}-2c_{\mp}\exp(\pi\Delta\theta)+2c_{\pm}\exp(-\pi\Delta\theta)\right\},\qquad(6.3.19)$$

where

$$c_{\pm}=\frac{\exp\{-2\pi(\pi\pm\alpha)\Delta_{\mp}\mathrm{i}2\pi(\Lambda-\frac{1}{2})\}}{1-\exp\{-2\pi(\pi\pm\alpha)\Delta\mp\mathrm{i}2\pi(\Lambda-\frac{1}{2})\}}.\qquad(6.3.20)$$

Hence we obtain

$$\bar{f}_{\mathrm{SA}}(\theta)=\frac{\pi\Delta\theta}{\sinh(\pi\Delta\theta)}\cosh(\alpha\Delta\theta)-2\pi\Delta\theta\times[c_{-}\sinh\{(\pi-\alpha)\Delta\theta\}+c_{+}\sinh\{(\pi+\alpha)\Delta\theta\}]\qquad(6.3.21)$$

$$\tilde{f}_{\mathrm{SA}}(\theta)=\frac{\pi\Delta\theta}{\sinh(\pi\Delta\theta)}\frac{\sinh(\alpha\Delta\theta)}{\theta}+2\pi\Delta\times[c_{-}\cosh\{(\pi-\alpha)\Delta\theta\}-c_{+}\cosh\{(\pi+\alpha)\Delta\theta\}].\qquad(6.3.22)$$

In these expressions, the first terms are the $m=0$ contributions while the rest comes from the sum of the $m\neq 0$ terms. Equation (6.3.22) shows that $\tilde{f}_{\mathrm{SA}}(\theta)$ vanishes if the phase parameter α approaches zero. For $\alpha=0$, the relative magnitude of the $m\neq 0$ and $m=0$ contributions to $\bar{f}_{\mathrm{SA}}(\theta)$ is given by (assuming for simplicity that $\Lambda-\frac{1}{2}=l_0$ is an integer)

$$4c\{\sinh(\pi\Delta\theta)\}^2\approx 4\{\sinh(\pi\Delta\theta)\}^2\exp(-2\pi^2\Delta).\qquad(6.3.23)$$

This ratio is zero at $\theta=0$ and small everywhere else except for very small Δ and θ close to π. Thus, except in the idealized sharp-cut-off limit $\Delta\to 0$, we are justified in neglecting all but the $m=0$ term in the Poisson series.

Then we can write eqn (6.3.12) as

$$\begin{aligned}f(\theta)&=\frac{\Lambda}{2k}\frac{F(\Delta\theta)}{(\theta\sin\theta)^{1/2}}[\exp(\alpha\Delta\theta)\{J_0(\Lambda\theta)+\mathrm{i}J_1(\Lambda\theta)\}-\exp(-\alpha\Delta\theta)\\&\quad\times\{J_0(\Lambda\theta)-\mathrm{i}J_1(\Lambda\theta)\}]\\&\simeq\left(\frac{\Lambda}{2\pi\sin\theta}\right)^{1/2}\frac{F(\Delta\theta)}{k\theta}[\exp(\alpha\Delta\theta)\exp\{\mathrm{i}(\Lambda\theta-\tfrac{1}{4}\pi)\}\\&\quad-\exp(-\alpha\Delta\theta)\exp\{-\mathrm{i}(\Lambda\theta-\tfrac{1}{4}\pi)\}]\end{aligned}\qquad(6.3.24)$$

and eqn (6.3.16) as

$$f(\theta)=\frac{\Lambda^2}{2k}\left\{\mathrm{i}\cosh(\alpha\Delta\theta)\frac{2J_1(\Lambda\theta)}{\Lambda\theta}+\frac{2\sinh(\alpha\Delta\theta)}{\Lambda\theta}J_0(\Lambda\theta)\right\}\left(\frac{\theta}{\sin\theta}\right)^{1/2}F(\Delta\theta).\qquad(6.3.25)$$

Note that for a small phase parameter α this reduces to

$$f(\theta) \approx \frac{\Lambda^2}{2k}\left\{\mathrm{i}\frac{2J_1(\Lambda\theta)}{\Lambda\theta} + \frac{2\alpha\Delta}{\Lambda}J_0(\Lambda\theta)\right\}\left(\frac{\theta}{\sin\theta}\right)^{1/2} F(\Delta\theta), \qquad (6.3.26)$$

which is identical with the SAM (Strong Absorption Model) expression given by Frahn and Venter (1963 [6.5]) (with $\alpha\Delta = \mu$).

These expressions show, first, the overall damping by the function $F(\Delta\theta)$, which decreases rapidly with increasing $\Delta\theta$ from $F(0) = 1$. The second part of eqn (6.3.24) displays most clearly the effect of the phase parameter α: it causes a disparity in the relative magnitude of the (+) and (−) branches, the near-side branch being smaller than the far-side branch by a factor $\exp(-2\alpha\Delta\theta)$. Thus 'refractive damping' in the Fraunhofer diffraction pattern has been discussed by Kauffmann (1977 [6.6]) and Hartmann (1976 [6.7]) and is borne out by experimental data for alpha scattering.

Going back to the general expression (6.3.16), we note that $\bar{f}_{\mathrm{SA}}(0) = 1$, and we obtain for the forward scattering amplitude

$$f(0) = \mathrm{i}\frac{\Lambda^2}{2k} + \frac{\Lambda}{k}\tilde{f}_{\mathrm{SA}}(0), \qquad (6.3.27)$$

where

$$\tilde{f}_{\mathrm{SA}}(0) = \Delta F'_{\mathrm{N}}(0) + \sum_{m\neq 0} \frac{F_{\mathrm{N}}(2m\pi\Delta)}{2m\pi}\exp\{\mathrm{i}2m\pi(\Lambda - \tfrac{1}{2})\}$$
$$\sim \alpha\Delta + 2\pi\Delta(c_- - c_+), \qquad (6.3.28)$$

with the last expression obtained from (6.3.22). Then we find from the optical theorem for the total cross-section

$$\sigma_{\mathrm{tot}} = \frac{4\pi}{k}\,\mathrm{Im}\, f(0) = 2\pi\left(\frac{\Lambda}{k}\right)^2\left\{1 + \frac{2}{\Lambda}\,\mathrm{Im}\,\tilde{f}_{\mathrm{SA}}(0)\right\} \approx 2\pi\left(\frac{\Lambda}{k}\right)^2, \qquad (6.3.29)$$

which is in agreement with the 'geometrical' value $2\pi R^2$ if $\Lambda = kR$.

Now we turn to large-angle scattering, using the asymptotic form (6.3.6) in the Poisson representation (6.3.3). Then a similar calculation to that which led to eqn (6.3.12) yields

$$f(\theta) = \frac{\Lambda}{2k}\left(\frac{\vartheta}{\sin\vartheta}\right)^{1/2}$$
$$\times[G^{(0)}_{\mathrm{LA}}(\vartheta)\{J_0(\Lambda\vartheta) - \mathrm{i}J_1(\Lambda\vartheta)\} + G^{(0)}_{\mathrm{LA}}(-\vartheta)\{J_0(\Lambda\vartheta) + \mathrm{i}J_1(\Lambda\vartheta)\}], \qquad (6.3.30)$$

where $\vartheta = \pi - \theta$, and

$$G^{(0)}_{\mathrm{LA}}(\pm\vartheta) = \sum_{m=-\infty}^{\infty} \exp\{\mathrm{i}(2m+1)\pi(\Lambda - \tfrac{1}{2})\}\frac{F_{\mathrm{N}}\{\Delta\langle(2m+1)\pi \mp \vartheta\rangle\}}{(2m-1)\pi \mp \vartheta}, \qquad (6.3.31)$$

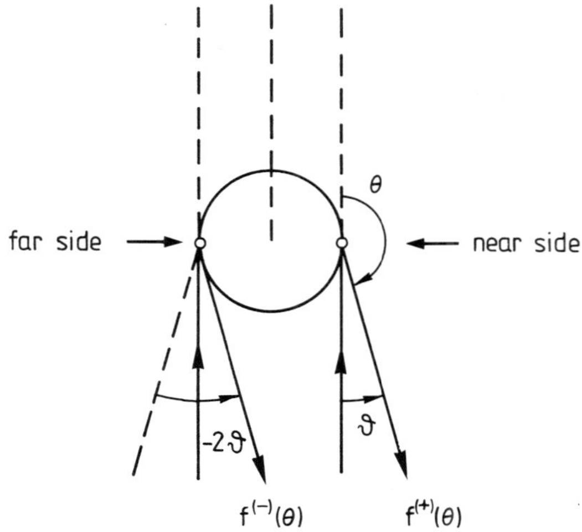

FIG. 6.3. Near-side and far-side scattering to backward angles.

with the subscript LA standing for large angle. We note the following relation between the functions $G^{(0)}_{\mathrm{LA}}$ and $G^{(0)}_{\mathrm{SA}}$ defined by (6.3.13):

$$G^{(0)}_{\mathrm{LA}}(\pm\vartheta) = \exp\{\pm \mathrm{i}\pi(\Lambda - \tfrac{1}{2})\} G^{(0)}_{\mathrm{SA}}(\mp\theta). \tag{6.3.32}$$

Again the structure of eqn (6.3.30) is transparent: $f(\theta)$ consists of two branches corresponding to scattering from the near side and the far side of the nucleus, as shown in Fig. 6.3. Using the asymptotic expressions (6.3.14), we obtain

$$f(\theta) \simeq \frac{1}{k}\left(\frac{\Lambda}{2\pi \sin\vartheta}\right)^{1/2}$$

$$[G^{(0)}_{\mathrm{LA}}(\vartheta) \exp\{-\mathrm{i}(\Lambda\vartheta - \tfrac{1}{4}\pi)\} + G^{(0)}_{\mathrm{LA}}(-\vartheta)\exp\{\mathrm{i}(\Lambda\vartheta - \tfrac{1}{4}\pi)\}], \tag{6.3.33}$$

But by virtue of relation (6.3.32) this is equal to the form

$$f(\theta) \simeq \frac{1}{k}\left(\frac{\Lambda}{2\pi \sin\theta}\right)^{1/2} [G^{(0)}_{\mathrm{SA}}(\theta)\exp\{-\mathrm{i}(\Lambda\theta - \tfrac{1}{4}\pi)\} + G^{(0)}_{\mathrm{SA}}(-\theta)\exp\{\mathrm{i}(\Lambda\theta - \tfrac{1}{4}\pi)\}], \tag{6.3.34}$$

which is obtained by using the asymptotics (6.3.14) in eqn (6.3.12). The agreement between the two expressions (6.3.33) and (6.3.34) for intermediate angles shows that the formulae (6.3.12) and (6.3.30) together give a *uniform* representation of the scattering amplitude over the whole angular range from $\theta = 0$ to $\theta = \pi$, as shown in Fig. 6.4.

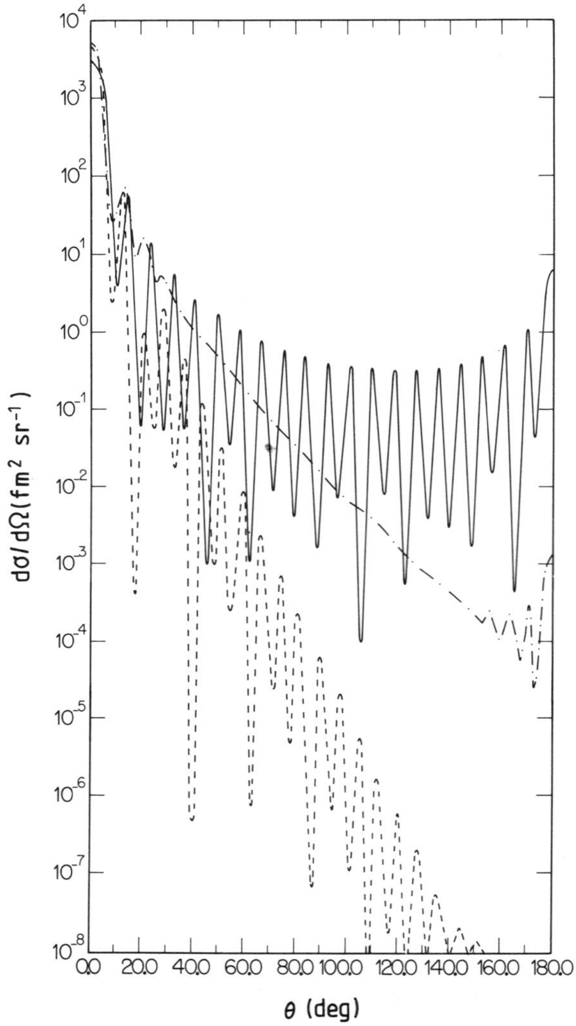

FIG. 6.4. Differential cross-section for the 'neutral' scattering of $^{16}O+^{12}C$ (zero Sommerfeld parameter) at a centre-of-mass energy $E_{cm} = 50$ MeV: ——, sharp-cut-off limit $\Delta = 0$; – – –, $\Delta = 1.5$, $\alpha = 0$; –·–, $\Delta = 1.5$, $\alpha = 1.57$.

By introducing the functions

$$\bar{f}_{\mathrm{LA}}(\theta)=\tfrac{1}{2}\{G_{\mathrm{LA}}^{(0)}(\vartheta)+G_{\mathrm{LA}}^{(0)}(-\vartheta)\}$$
$$\tilde{f}_{\mathrm{LA}}(\theta)=\tfrac{1}{2}\{G_{\mathrm{LA}}^{(0)}(\vartheta)-G_{\mathrm{LA}}^{(0)}(-\vartheta)\} \tag{6.3.35}$$

eqn (6.3.30) can be written in the alternative form

$$f(\theta)=\frac{\Lambda}{k}\left(\frac{\pi-\theta}{\sin\theta}\right)^{1/2}[\bar{f}_{\mathrm{LA}}(\theta)J_0\{\Lambda(\pi-\theta)\}-\mathrm{i}\tilde{f}_{\mathrm{LA}}(\theta)J_1\{\Lambda(\pi-\theta)\}]. \tag{6.3.36}$$

From the expressions (6.3.31) it is seen that at large angles there are *two* dominant Poisson terms, with $m=0$ and $m=1$, and we have to a good approximation

$$G_{\mathrm{LA}}^{(0)}(\vartheta)\approx\frac{F_{\mathrm{N}}\{\Delta(\pi-\vartheta)\}}{\pi-\vartheta}\exp\{\mathrm{i}\pi(\Lambda-\tfrac{1}{2})\}-\frac{F_{\mathrm{N}}\{-\Delta(\pi+\vartheta)\}}{\pi+\vartheta}\exp\{-\mathrm{i}\pi(\Lambda-\tfrac{1}{2})\} \tag{6.3.37}$$

$$G_{\mathrm{LA}}^{(0)}(-\vartheta)\approx\frac{F_{\mathrm{N}}\{\Delta(\pi+\vartheta)\}}{\pi+\vartheta}\exp\{\mathrm{i}\pi(\Lambda-\tfrac{1}{2})\}-\frac{F_{\mathrm{N}}[-\Delta(\pi-\vartheta)]}{\pi-\vartheta}\exp\{-\mathrm{i}\pi(\Lambda-\tfrac{1}{2})\}.$$

In the backward direction $\theta=\pi$ $(\vartheta=0)$ we obtain

$$\begin{aligned} f(\pi)&=\frac{\Delta}{k}G_{\mathrm{LA}}^{(0)}(0)=\frac{\Lambda}{k}\bar{f}_{\mathrm{LA}}(\pi)\\ &\approx\frac{\Lambda}{\pi k}[F_{\mathrm{N}}(\Delta\pi)\exp\{\mathrm{i}\pi(\Lambda-\tfrac{1}{2})\}-F_{\mathrm{N}}(-\Delta\pi)\exp\{-\mathrm{i}\pi(\Lambda-\tfrac{1}{2})\}]. \end{aligned} \tag{6.3.38}$$

When the form (6.3.17) is used this becomes

$$f(\pi)\approx\frac{2\Lambda}{\pi k}\{\sinh(\alpha\Delta\pi)\sin(\pi\Lambda)-\mathrm{i}\cosh(\alpha\Delta\pi)\cos(\pi\Lambda)\}F(\Delta\pi). \tag{6.3.39}$$

From this and eqn (6.3.27) we can calculate the ratio of the backward to forward scattering cross-section as

$$\begin{aligned} \frac{\sigma(\pi)}{\sigma(0)}&\approx\left(\frac{4}{\pi\Lambda}\right)^2\{\cos^2(\pi\Lambda)+\sinh^2(\alpha\Delta\pi)\}\{F(\Delta\pi)\}^2\\ &\simeq\{\cos^2(\pi\Lambda)+\sinh^2(\alpha\Delta\pi)\}\left(\frac{8\pi\Delta}{\Lambda}\right)^2\exp(-2\pi^2\Delta). \end{aligned} \tag{6.3.40}$$

This ratio is very small under all conditions.

6.4. Charged particles

We now include the Rutherford phase shifts σ_l in the total S-matrix $S_l = S_{N,l}\exp(i2\sigma_l)$, where

$$\sigma_l = \arg\Gamma(l+1+in) \tag{6.4.1}$$

and n is the Sommerfeld parameter. Coulomb interaction always dominates in elastic scattering at small angles because the Rutherford scattering amplitude

$$\begin{aligned} f_R(\theta) &= -\frac{i}{k}\sum_{l=0}^{\infty}(l+\tfrac{1}{2})\exp(i2\sigma_l)P_l(\cos\theta) \\ &= -\frac{n}{2k}\frac{1}{(\sin\frac{1}{2}\theta)^2}\exp[i\{2\sigma_0 - 2n\ln(\sin\tfrac{1}{2}\theta)\}] \end{aligned} \tag{6.4.2}$$

is singular at $\theta = 0$. Comparison of eqn (6.4.2) with eqn (6.3.16) shows that $f_R(\theta)$ dominates over the neutral amplitude for $\theta \lesssim 2n^{1/2}/\Lambda$. Thus, if $n \gtrsim \frac{1}{4}$ we can replace $P_l(\cos\theta)$ by either of the asymptotic forms (6.3.6) or (6.3.7). Further we use the result of the previous section that the $m=0$ term in the Poisson series is dominant in strong-absorption scattering at all angles except close to 180°. Since large-angle scattering will be treated separately in Section 6.8, we can confine ourselves here to the $m=0$ contribution and use the asymptotics (6.3.7). Then the amplitude can be split into near-side and far-side branches:

$$f(\theta) = f^{(+)}(\theta) + f^{(-)}(\theta), \tag{6.4.3}$$

where

$$f^{(\pm)}(\theta) = -\frac{i}{k}\frac{1}{(2\pi\sin\theta)^{1/2}}\int_0^{\infty} d\lambda\,\lambda^{1/2}S_N(\lambda)\exp\{i\phi_\pm(\lambda,\theta)\} \tag{6.4.4}$$

with

$$\phi_\pm(\lambda,\theta) = 2\sigma(\lambda) \mp (\lambda\theta - \tfrac{1}{4}\pi). \tag{6.4.5}$$

Our next step is to introduce an integral representation of $S_N(\lambda)$,

$$S_N(\lambda) = \frac{i}{2\pi}\lim_{\varepsilon\to+0}\int_{-\infty}^{\infty} dz\,\frac{\exp\{-i(\lambda-\Lambda)z\}}{z+i\varepsilon}F_N(\Delta z), \tag{6.4.6}$$

in terms of the Fourier transform (6.3.11) of the derivative $D_N(\lambda) = dS_N(\lambda)/d\lambda$. Then eqns (6.4.4) can be written in the form (Frahn and Gross 1976 [6.8])

$$f^{(+)}(\theta) = \frac{i}{2\pi}\lim_{\varepsilon\to+0}\int_{-\infty}^{\infty} dz\,\frac{F_N(\Delta z)\exp(i\Lambda z)}{z+i\varepsilon}\left\{\frac{\sin(z+\theta)}{\sin\theta}\right\}^{1/2} f_0^{(+)}(z+\theta) \tag{6.4.7}$$

$$f^{(-)}(\theta)=\frac{1}{2\pi}\lim_{\varepsilon\to+0}\int_{-\infty}^{\infty}\mathrm{d}z\,\frac{F_{\mathrm{N}}(\Delta z)\exp(\mathrm{i}\Lambda z)}{z+\mathrm{i}\varepsilon}\left\{\frac{\sin(z-\theta)}{\sin\theta}\right\}^{1/2}f_0^{(+)}(z-\theta), \tag{6.4.8}$$

where

$$f_0^{(+)}(\vartheta)=-\frac{\mathrm{i}}{k}\frac{1}{(2\pi\sin\vartheta)^{1/2}}\int_0^{\infty}\mathrm{d}\lambda\lambda^{1/2}\exp\{\mathrm{i}\phi_+(\lambda,\vartheta)\}. \tag{6.4.9}$$

The integral in (6.4.9) receives its main contribution from the point of stationary phase λ_ϑ defined by

$$\Theta_{\mathrm{R}}(\lambda_\vartheta)=\vartheta, \qquad \lambda_\vartheta=\Theta_{\mathrm{R}}^{-1}(\vartheta), \tag{6.4.10}$$

where $\Theta_{\mathrm{R}}(\lambda)=\mathrm{d}\{2\sigma(\lambda)\}/\mathrm{d}\lambda$ is the quantal deflection function for Rutherford scattering. Evaluation of eqn (6.4.9) in the stationary phase approximation (SPA) yields

$$f_0^{(+)}(\vartheta)=-\frac{\mathrm{i}}{k}\left\{\frac{\lambda_\vartheta}{-\Theta_{\mathrm{R}}'(\lambda_\vartheta)\sin\vartheta}\right\}^{1/2}\exp[\mathrm{i}\{2\sigma(\lambda_\vartheta)-\lambda_\vartheta)-\lambda_\vartheta\vartheta\}]. \tag{6.4.11}$$

Consistent with our general assumptions we can use the asymptotic form of the Rutherford phase shift function for large λ and n:

$$2\sigma(\lambda)=2\sigma_0-2n\ln(\sin\tfrac{1}{2}\varepsilon)+n\varepsilon\cot\tfrac{1}{2}\varepsilon-\tfrac{1}{2}\pi+O\{(\lambda^2+n^2)^{-1/2}\} \tag{6.4.12}$$

where

$$\varepsilon=2\arctan(n/\lambda). \tag{6.4.13}$$

Then we have

$$\Theta_{\mathrm{R}}(\lambda)=2\arctan\left(\frac{n}{\lambda}\right)=\varepsilon, \qquad \Theta_{\mathrm{R}}'(\lambda)=-\frac{2n}{\lambda^2+n^2}=-\frac{2}{n}(\sin\tfrac{1}{2}\varepsilon)^2, \tag{6.4.14}$$

and from (6.4.10),

$$\lambda_\vartheta=n\cot\tfrac{1}{2}\vartheta \tag{6.4.15}$$

$$2\sigma(\lambda_\vartheta)=2\sigma_0-2n\ln(\sin\tfrac{1}{2}\vartheta)+\lambda_\vartheta\vartheta-\tfrac{1}{2}\pi \tag{6.4.16}$$

$$\Theta_{\mathrm{R}}'(\lambda_\vartheta)=-\frac{2}{n}(\sin\tfrac{1}{2}\vartheta)^2. \tag{6.4.17}$$

Thus eqn (6.4.11) becomes

$$f_0^{(+)}(\vartheta)=-\frac{n}{2k}\frac{1}{(\sin\tfrac{1}{2}\vartheta)^2}\exp[\mathrm{i}\{2\sigma_0-2n\ln(\sin\tfrac{1}{2}\vartheta)\}]=f_{\mathrm{R}}(\vartheta). \tag{6.4.18}$$

(It is one of the remarkable features of Coulomb scattering that the SPA together with the asymptotic phase shifts yields the *exact* form of the Rutherford scattering amplitude. Note that the negative branch $f_0^{(-)}(\vartheta)$ gives no contribution in the SPA.)

The reason for introducing the representation (6.4.6) is that, since $S_N(\lambda)$ is a rapidly varying function in the case of strong absorption, the Fourier transform $F_N(\Delta z)$ varies slowly and the resulting integration over angles can be carried out in SPA. Inserting the SPA expression (6.4.11) in eqns (6.4.7) and (6.4.8) gives

$$f^{(+)}(\theta)=\frac{1}{2\pi k}\frac{\exp(-\mathrm{i}\Lambda\theta)}{(\sin\theta)^{1/2}}\lim_{\varepsilon\to+0}\int_0^{\pi}\mathrm{d}\vartheta\left\{\frac{\lambda_\vartheta}{-\Theta_R'(\lambda_\vartheta)}\right\}^{1/2}\times\frac{F_N[\Delta(\vartheta-\theta)]}{\nu-\theta+\mathrm{i}\epsilon}\exp[\mathrm{i}\{2\sigma(\lambda_\vartheta)-(\lambda_\vartheta-\Lambda)\vartheta\}]$$

$$f^{(-)}(\theta)=-\frac{\mathrm{i}}{2\pi k}\frac{\exp(\mathrm{i}\Lambda\theta)}{(\sin\theta)^{1/2}}\lim_{\varepsilon\to+0}\int_0^{\pi}\mathrm{d}\vartheta\left\{\frac{\lambda_\vartheta}{-\Theta_R'(\lambda_\vartheta)}\right\}^{1/2}\times\frac{F_N\{\Delta(\vartheta+\theta)\}}{\vartheta+\theta+\mathrm{i}\epsilon}\exp[\mathrm{i}\{2\sigma(\lambda_\vartheta)-(\lambda_\vartheta-\Lambda)\vartheta\}], \quad (6.4.19)$$

where λ_ϑ is the stationary point pertaining to the angle $\vartheta=z+\theta$ in eqn (6.4.19) and to $\vartheta=z-\theta$ in eqn (6.4.20). The range of the integration over ϑ is restricted to physical angles because the SPA expression (6.4.11) is defined only in this range and zero outside.

The oscillatory parts of the integrands in (6.4.19) and (6.4.20) have a point of stationary phase ϑ_s, which in both branches is given by

$$\vartheta_s=\Theta_R(\Lambda)\equiv\theta_R=2\arctan(n/\Lambda), \quad (6.4.21)$$

i.e. by the 'Rutherford grazing angle'. We now expand the slowly varying functions $F_N\{\Delta(\vartheta\pm\theta)\}$ about this point to all orders:

$$\begin{aligned}F_N\{\Delta(\vartheta\mp\theta)\}&=\sum_{p=0}^{\infty}\frac{1}{p!}\{\Delta(\vartheta-\theta_R)\}^p F_N^{(p)}\{\Delta(\theta_R\mp\theta)\}\\&=\sum_{p=0}^{\infty}\frac{\Delta^p}{p!}F_N^{(p)}\{\Delta(\theta_R\mp\theta)\}\{[\vartheta\mp\theta)-(\theta_R\mp\theta)\}^p\\&=\sum_{p=0}^{\infty}\frac{\Delta^p}{p!}F_N^{(p)}\{\Delta(\theta_R\mp\theta)\}\sum_{q=0}^{p}(-)^{p-q}\binom{p}{q}(\vartheta\mp\theta)^q(\theta_R\mp\theta)^{p-q}.\end{aligned} \quad (6.4.22)$$

Because of the poles near $\vartheta=\theta$ and $\vartheta=-\theta$ in the integrands in (6.4.19) and (6.4.20) respectively, we separate the terms with $q=0$ from the sums in (6.4.22). This yields

$$f_{q=0}^{(\pm)}(\theta)=F_{\mathrm{N}}(0)f_{\mathrm{sco}}^{(\pm)}(\theta)=f_{\mathrm{sco}}^{(\pm)}(\theta), \tag{6.4.23}$$

since $F_{\mathrm{N}}(0)=1$, where

$$f_{\mathrm{sco}}^{(\pm)}(\theta)=\binom{1}{-\mathrm{i}}\frac{\exp(\mp\mathrm{i}\Lambda\theta)}{2\pi k(\sin\theta)^{1/2}}\lim_{\epsilon\to+0}\int_0^{\pi}\mathrm{d}\vartheta\left\{\frac{\lambda_\vartheta}{-\Theta_{\mathrm{R}}'(\lambda_\vartheta)}\right\}^{1/2} \times\frac{\exp[\mathrm{i}\{2\sigma(\lambda_\vartheta)-(\lambda_\vartheta-\Lambda)\vartheta\}]}{\vartheta\mp\theta+\mathrm{i}\epsilon} \tag{6.4.24}$$

are the two branches of the sharp-cut-off amplitude $f_{\mathrm{sco}}(\theta)=f_{\mathrm{sco}}^{(+)}(\theta)+f_{\mathrm{sco}}^{(-)}(\theta)$. The latter is so called because it is the amplitude for Rutherford scattering with $\Delta=0$, for which the reflection function

$$\eta_{\mathrm{sco}}(\lambda)=\frac{\mathrm{i}}{2\pi}\lim_{\epsilon\to+0}\int_{-\infty}^{\infty}\mathrm{d}z\frac{\exp\{-\mathrm{i}(\lambda-\Lambda)z\}}{z+\mathrm{i}\epsilon}=\begin{cases}0, & \lambda<\Lambda\\ 1, & \lambda\geqslant\Lambda\end{cases} \tag{6.4.25}$$

is a unit step function at $\lambda=\Lambda$, and the absorptive shape function is a delta function $D_{\mathrm{N}}(\lambda)=\delta(\lambda-\Lambda)$. After substituting $\vartheta=\Theta_{\mathrm{R}}(\lambda_\vartheta)$ in eqns (6.4.24) and using (6.4.25) we obtain

$$\begin{aligned} f_{\mathrm{sco}}^{(\pm)}(\theta)&=-\frac{\mathrm{i}}{k}\frac{1}{(2\pi\sin\theta)^{1/2}}\int_0^{\infty}\mathrm{d}\lambda\lambda^{1/2}\eta_{\mathrm{sco}}(\lambda)\exp\{\mathrm{i}\phi_{\pm}(\lambda,\theta)\}\\ &=-\frac{\mathrm{i}}{k}\frac{1}{(2\pi\sin\theta)^{1/2}}\int_{\Lambda}^{\infty}\mathrm{d}\lambda\lambda^{1/2}\exp\{\mathrm{i}\phi_{\pm}(\lambda,\theta)\}. \end{aligned} \tag{6.4.26}$$

The remaining terms with $q>0$ in the expansion (6.4.22) give no contributions from the poles near $\vartheta=\pm\theta$; we have

$$f_{q>0}^{(\pm)}(\theta)=\sum_{p=1}^{\infty}\frac{\Delta^p}{p!}F_{\mathrm{N}}^{(p)}\{\Delta(\theta_{\mathrm{R}}\mp\theta)\}\sum_{q=1}^{p}(-)^{p-q}\binom{p}{q}(\theta_{\mathrm{R}}\mp\theta)^{p-q}g_{q-1}^{(\pm)}(\theta), \tag{6.4.27}$$

where

$$\begin{aligned} g_{q-1}^{(\pm)}(\theta)&=\binom{1}{-\mathrm{i}}\frac{\exp(\mp\mathrm{i}\Lambda\theta)}{2\pi k(\sin\theta)^{1/2}}\int_0^{\pi}\mathrm{d}\vartheta\left\{\frac{\lambda_\vartheta}{-\Theta_{\mathrm{R}}'(\lambda_\vartheta)}\right\}^{1/2}(\vartheta\mp\theta)^{q-1}\\ &\quad\times\exp[\mathrm{i}\{2\sigma(\lambda_\vartheta)-(\lambda_\vartheta-\Lambda)\vartheta\}]\\ &=\binom{1}{-\mathrm{i}}\frac{\exp(\mp\mathrm{i}\Lambda\theta)}{2\pi k(\sin\theta)^{1/2}}\int_0^{\infty}\mathrm{d}\lambda\lambda^{1/2}\{-\Theta_{\mathrm{R}}'(\lambda)\}^{1/2}\{\Theta_{\mathrm{R}}(\lambda)\mp\theta\}^{q-1}\\ &\quad\times\exp[\mathrm{i}\{2\sigma(\lambda)-(\lambda-\Lambda)\theta_{\mathrm{R}}(\lambda)\}] \end{aligned} \tag{6.4.28}$$

after the substitution $\vartheta=\Theta_{\mathrm{R}}(\lambda)$. The integrals in the second part of eqn

(6.4.28) can be evaluated in SPA with the result

$$g_{q-1}^{(\pm)}(\theta)=\frac{\kappa(\theta)}{k}(\theta_{\mathrm{R}}\mp\theta)^{q-1}\exp\{\mathrm{i}\phi_{\pm}(\Lambda,\theta)\} \tag{6.4.29}$$

where

$$\kappa(\theta)=\left(\frac{\Lambda}{2\pi\sin\theta}\right)^{1/2}. \tag{6.4.30}$$

After inserting (6.4.29) in (6.4.27) the sums over q and p can be performed, and we obtain, together with eqn (6.4.23),

$$f^{(\pm)}(\theta)=f_{\mathrm{sco}}^{(\pm)}(\theta)+\frac{\kappa(\theta)}{k}\frac{F_{\mathrm{N}}\{\Delta(\theta_{\mathrm{R}}\mp\theta)\}-1}{\theta_{\mathrm{R}}\mp\theta}\exp\{\mathrm{i}\phi_{\pm}(\Lambda,\theta\}. \tag{6.4.31}$$

By deriving this expression we have reduced the calculation of $f(\theta)$ to the evaluation of the sharp-cut-off amplitude $f_{\mathrm{sco}}(\theta)$. Since the latter is determined solely by the value of Λ (for a given Sommerfeld parameter n), our result (6.4.31) has the significant feature that *the details of the nuclear interaction in strong-absorption scattering enter only through the Fourier transforms* $F_{\mathrm{N}}\{\Delta(\theta_{\mathrm{R}}\mp\theta)\}$. (Note that eqn (6.4.31) generalizes earlier expressions (Frahn 1975 [6.9]; Frahn and Gross 1976 [6.8]), in which only the lowest-order terms in the expansion (6.4.22) were retained.)

The sharp-cut-off amplitude $f_{\mathrm{sco}}(\theta)$ can be evaluated, in either the form (6.4.24) or the form (6.4.26), by a systematic method that leads to an asymptotic expansion in descending orders of the large parameter Λ. Since this method has been described in detail (Frahn 1975 [6.9]; Frahn and Gross 1976 [6.8]), only the result is given here:

$$f_{\mathrm{sco}}^{(+)}(\theta)=\begin{cases} f_{\mathrm{R}}(\theta)+\dfrac{\kappa(\theta)}{k}\Gamma_{<}(\theta_{\mathrm{R}}-\theta)\exp\{\mathrm{i}\phi_{+}(\Lambda,\theta)\}, & \theta\leqslant\theta_{\mathrm{R}} \quad (6.4.32a)\\[2ex] \dfrac{\kappa(\theta)}{k}\Gamma_{>}(\theta_{\mathrm{R}}-\theta)\exp\{\mathrm{i}\phi_{+}(\Lambda,\theta)\}, & \theta\geqslant\theta_{\mathrm{R}} \quad (6.4.32b)\end{cases}$$

$$f_{\mathrm{sco}}^{(-)}(\theta)=\frac{\kappa(\theta)}{k}\frac{\exp\{\mathrm{i}\phi_{-}(\Lambda,\theta)\}}{\theta_{\mathrm{R}}+\theta}. \tag{6.4.32c}$$

The functions $\Gamma_{\lessgtr}(\theta_{\mathrm{R}}-\theta)$ are defined as

$$\Gamma_{\lessgtr}(\theta_{\mathrm{R}}-\theta)=\pm\gamma(\mp u)\{1+(\theta_{\mathrm{R}}-\theta)a_1(u)\}-a_0(u), \tag{6.4.33}$$

where

$$\gamma(u)=\tfrac{1}{2}\operatorname{erfc}\{\exp(\mathrm{i}\tfrac{1}{4}\pi)u\}2\pi\kappa(\theta_{\mathrm{R}})\exp\{\mathrm{i}(u^2+\tfrac{1}{4}\pi)\} \tag{6.4.34}$$

with the complementary error function

$$\operatorname{erfc}(z)=\frac{2}{\pi^{1/2}}\int_z^\infty \mathrm{d}\tau \exp(-\tau^2) \tag{6.4.35}$$

of argument

$$u=\left(\frac{\Lambda}{2\sin\theta_\mathrm{R}}\right)^{1/2}(\theta-\theta_\mathrm{R})=\frac{n^{1/2}}{2\sin\frac{1}{2}\theta_\mathrm{R}}(\theta-\theta_\mathrm{R}). \tag{6.4.36}$$

The functions $a_0(u)$ and $a_1(u)$ are defined by

$$a_0(u)=\frac{1}{2\sin\theta_\mathrm{R}}\{1+\tfrac{4}{3}(1+\mathrm{i}u^2)(\cos\tfrac{1}{2}\theta_\mathrm{R})^2\} \tag{6.4.37}$$

$$a_1(u)=\frac{1}{2\sin\theta_\mathrm{R}}\{1+2(1+\mathrm{i}\tfrac{2}{3}u^2)(\cos\tfrac{1}{2}\theta_\mathrm{R})^2\}. \tag{6.4.38}$$

The reason for $f_{\mathrm{sco}}^{(+)}(\theta)$ having different functional forms in the regions $\theta\leqslant\theta_\mathrm{R}$ and $\theta\geqslant\theta_\mathrm{R}$ is that the integral in eqn (6.4.24) for $f_{\mathrm{sco}}^{(+)}(\theta)$ receives its main contributions not only from the point of stationary phase, $\vartheta_\mathrm{s}=\theta_\mathrm{R}$, but also from the pole $\vartheta_\mathrm{p}=\theta$ which coincides with ϑ_s at $\theta=\theta_\mathrm{R}$. Alternatively, the main contributions to the integral in eqn (6.4.26) come from both the stationary point $\lambda_\mathrm{s}=\Theta_\mathrm{R}^{-1}(\theta)$ and from the lower integration limit $\lambda=\Lambda$, which coincides with λ_s for $\theta=\theta_\mathrm{R}$. In contrast, the pole $\vartheta_\mathrm{p}=-\theta$ in the integrand in eqn (6.4.24) for $f_{\mathrm{sco}}^{(-)}(\theta)$ lies outside the range of integration and the integral receives its main contribution only from the stationary point $\vartheta_\mathrm{s}=\theta_\mathrm{R}$; alternatively, there is no stationary point within the range of the λ integration in eqn (6.4.26) for $f_{\mathrm{sco}}^{(-)}(\theta)$, so that the main contribution comes only from the cut-off point $\lambda=\Lambda$. As will be shown in the next section, the two expressions (6.4.32*a*) and (6.4.32*b*) are differentiably continuous at $\theta=\theta_\mathrm{R}$.

Equations (6.4.31) and (6.4.32) together give our final closed-form result for the elastic scattering amplitude:

$$f^{(+)}(\theta)=f_\mathrm{R}(\theta)+\frac{\kappa(\theta)}{k}\left[\Gamma_<(\theta_\mathrm{R}-\theta)+\frac{F_\mathrm{N}\{\Delta(\theta_\mathrm{R}-\theta)\}-1}{\theta_\mathrm{R}-\theta}\right]\exp\{\mathrm{i}\phi_+(\Lambda,\theta)\},$$

$$\theta\leqslant\theta_\mathrm{R} \tag{6.4.39a}$$

$$f^{(+)}(\theta)=\frac{\kappa(\theta)}{k}\left[\Gamma_>(\theta_\mathrm{R}-\theta)+\frac{F_\mathrm{N}\{\Delta(\theta_\mathrm{R}-\theta)\}-1}{\theta_\mathrm{R}-\theta}\right]\exp\{\mathrm{i}\phi_+(\Lambda,\theta)\},$$

$$\theta>\theta_\mathrm{R} \tag{6.4.39b}$$

$$f^{(-)}(\theta)=\frac{\kappa(\theta)}{k}\frac{F_\mathrm{N}\{\Delta(\theta_\mathrm{R}+\theta)\}}{\theta_\mathrm{R}+\theta}\exp\{\mathrm{i}\phi_-(\Lambda,\theta)\}. \tag{6.4.39c}$$

These expressions comprise all the leading contributions, up to terms of order Λ^{-1} (or of order n^{-1} for fixed θ_R), in an asymptotic expansion of the partial-wave series of the amplitude for strong-absorption scattering.

6.5 Physical interpretation

To discuss the physical implications of the formulae (6.4.39) let us consider first the 'Coulomb' limit ($\mathscr{C}$-limit) and the 'neutral' limit ($\mathscr{N}$-limit) as defined in Section 6.3. As shown there, in the $\mathscr{C}$-limit, where $\Lambda \gg 1$ and $n \gg 1$ at fixed ratio n/Λ (i.e. fixed θ_R), the scattering is dominated by Fresnel-type diffraction. It applies to heavy-ion collisions at medium energies above the Coulomb barrier, which are characterized by both strong absorption and strong Coulomb interaction. On the other hand, in the $\mathscr{N}$-limit, where $\Lambda \gg 1$ but $n/\Lambda \ll 1$ (i.e. $\theta_R \ll \pi$), the diffractive scattering is predominantly of the Fraunhofer type. This applies to light-ion collisions at medium and higher energies, in the case of strong absorption but relatively weak Coulomb interaction. In the extreme $\mathscr{N}$-limit where $n \to 0$ ($\theta_R \to 0$), our expressions should of course be consistent with those derived in Section 6.3.

In the Fresnel regime it is convenient to discuss the ratio $f(\theta)/f_R(\theta)$ of the elastic amplitude to the Rutherford amplitude; indeed, experimental angular distributions for elastic heavy-ion scattering are normally presented as the ratio $\sigma(\theta)/\sigma_R(\theta)$. A little calculation shows that eqns (6.4.39) can be written in the form

$$\frac{f^{(+)}(\theta)}{f_R(\theta)} = 1 + A(\theta)\left[\Gamma_<(\theta_R - \theta) + \frac{F_N\{\Delta(\theta_R - \theta)\} - 1}{\theta_R - \theta}\right], \qquad \theta \leqslant \theta_R \qquad (6.5.1a)$$

$$\frac{f^{(+)}(\theta)}{f_R(\theta)} = A(\theta)\left[\Gamma_>(\theta_R - \theta) + \frac{F_N\{\Delta(\theta_R - \theta)\} - 1}{\theta_R - \theta}\right] \qquad \theta \geqslant \theta_R \qquad (6.5.1b)$$

$$\frac{f^{(-)}(\theta)}{f_R(\theta)} = -\mathrm{i}A(\theta)\frac{F_N\{\Delta(\theta_R + \theta)\}}{\theta_R + \theta}\exp(\mathrm{i}2\Lambda\theta) \qquad (6.5.1c)$$

where

$$A(\theta) = \frac{\kappa(\theta)}{k}\frac{\exp\{\mathrm{i}\phi_+(\Lambda, \theta)\}}{f_R(\theta)}$$
$$= -\frac{1}{2\pi\kappa(\theta)}\frac{\tan\frac{1}{2}\theta}{\tan\frac{1}{2}\theta_R}\exp\{-\mathrm{i}(u^2 + \tfrac{1}{4}\pi)\}\exp\{-\mathrm{i}\psi(u)\}. \qquad (6.5.2)$$

Here $\psi(u)$ is defined as the remainder of third and higher order in an expansion of the Rutherfore phase shift function $\sigma(\lambda_\theta)$ in powers of

$\theta-\theta_R$ or of the variable u:

$$2\sigma(\lambda_\theta)-(\lambda_\theta-\Lambda)\theta=2\sigma(\Lambda)+u^2+\psi(u). \tag{6.5.3}$$

Now we can explore the structure of the angular distribution, first in the 'illuminated region' ($\theta<\theta_R$), then in the 'shadow region' ($\theta>\theta_R$), and finally in the neighbourhood of the grazing angle ($\theta\approx\theta_R$).

6.5.1 *Illuminated region* $\theta<\theta_R$

The functions $\Gamma_{\gtrless}(\theta_R-\theta)$ have the simple asymptotic form

$$\Gamma_{\gtrless}(\theta_R-\theta)=\frac{1}{\theta_R-\theta}\left[1+\frac{i}{2\Lambda(\theta_R-\theta)}+O(\{\Lambda(\theta_R-\theta)\}^{-2})\right]. \tag{6.5.4}$$

Thus we obtain from eqns (6.5.1*a*) and (6.5.1*c*), for $\theta_R-\theta\gg(2\Lambda)^{-1}$,

$$\frac{f(\theta)}{f_R(\theta)}\simeq1+A(\theta)\left[\frac{F_N\{\Delta(\theta_R-\theta)\}}{\theta_R-\theta}-i\frac{F_N\{\Delta(\theta_R+\theta)\}}{\theta_R+\theta}\exp(i2\Lambda\theta)\right]. \tag{6.5.5}$$

By using eqn (6.5.2) this can be written in the form

$$\frac{f(\theta)}{f_R(\theta)}\simeq1+\frac{c(\theta)}{2\pi^{1/2}u}\exp\{-i(u^2+\tfrac{1}{4}\pi)\}\exp\{-i\psi(u)\}$$
$$\times F_N\{\Delta(\theta_R-\theta)\}\{1-ib(\theta)\exp(i2\Lambda\theta)\}, \tag{6.5.6}$$

where we have defined

$$c(\theta)=\left(\frac{\sin\theta}{\sin\theta_R}\right)^{1/2}\frac{\tan\frac{1}{2}\theta}{\tan\frac{1}{2}\theta_R},\qquad b(\theta)=\frac{F_N\{\Delta(\theta_R+\theta)\}}{\theta_R+\theta}\Bigg/\frac{F_N\{\Delta(\theta_R-\theta)\}}{\theta_R-\theta}. \tag{6.5.7}$$

Equation (6.5.6) can be interpreted as follows. The first two terms, from the near-side branch of $f(\theta)$, give rise to diffraction oscillations of the Fresnel type whose amplitudes are damped by the function $F_N\{\Delta(\theta_R-\theta)\}$. The third term, from the far-side branch of $f(\theta)$, causes a modulation by oscillations of Fraunhofer type, with angular period π/Λ, which are damped by the function $b(\theta)$.

To study the Fresnel-diffractive structure of the angular distribution, let us consider conditions of strong damping (which normally prevail in heavy-ion scattering). Then the cross-section ratio becomes

$$\frac{\sigma(\theta)}{\sigma_R(\theta)}\approx1-\frac{c(\theta)}{\pi^{1/2}|u|}\cos(u^2+\tfrac{1}{4}\pi)F_N\{\Delta(\theta_R-\theta)\},\qquad\theta\leqslant\theta_R-(2\Lambda)^{-1}. \tag{6.5.8}$$

The Fresnel oscillations have maxima when $u^2 = \frac{3}{4}\pi$, i.e. at the angles

$$\theta_m = \theta_R - \left\{\left(\tfrac{3}{4} + 2m\right)\frac{2\pi \sin\theta_R}{\Lambda}\right\}^{1/2}, \qquad m = 0, 1, 2, \ldots. \tag{6.5.9}$$

The main Fresnel peak, which is located at

$$\theta_0 = \theta_R - \left(\frac{3\pi \sin\theta_R}{2\Lambda}\right)^{1/2} = \theta_R - \left(\frac{3\pi}{n}\right)^{1/2} \sin\tfrac{1}{2}\theta_R, \tag{6.5.10}$$

has the magnitude

$$\left(\frac{\sigma}{\sigma_R}\right)_{\theta_0} \approx 1 + \frac{2}{\pi\sqrt{3}} c(\theta_0) F_N\{\Delta(\theta_R - \theta_0)\}. \tag{6.5.11}$$

If we use as an example the specific form (6.3.17) of $F_N(\Delta z)$, it is seen that the 'Coulomb damping' by the function $F\{\Delta(\theta_R - \theta_0)\}$ is counteracted by the factor $\exp\{\alpha\Delta(\theta_R - \theta_0)\}$ due to the nuclear phase. We shall give a pictorial explanation of these effects in Section 6.5.5.

6.5.2 *Shadow region* $\theta > \theta_R$

Using the asymptotic form (6.5.4) of $\Gamma_>(\theta_R - \theta)$, we obtain from eqns (6.5.1*b*) and (6.5.1*c*), for $\theta - \theta_R \gg (2\Lambda)^{-1}$,

$$\frac{f(\theta)}{f_R(\theta)} \simeq A(\theta)\left[\frac{F_N\{\Delta(\theta_R - \theta)\}}{\theta_R - \theta} - \mathrm{i}\frac{F_N\{\Delta(\theta_R + \theta)\}}{\theta_R + \theta}\exp(\mathrm{i}2\Lambda\theta)\right], \tag{6.5.12}$$

and the cross-section can be written as

$$\sigma(\theta) \simeq \frac{\Lambda}{2\pi k^2 \sin\theta}\left|\frac{F_N\{\Delta(\theta_R - \theta)\}}{\theta_R - \theta}\right|^2 \times |1 - ib(\theta)\exp(\mathrm{i}2\Lambda\theta)|^2, \qquad \theta \gtrsim \theta_R + (2\Lambda)^{-1}. \tag{6.5.13}$$

This shows that the elastic angular distribution falls off rapidly beyond the grazing angle θ_R owing to the strong (essentially exponential) decrease of the Fourier transform $F_N\{\Delta(\theta_R - \theta)\}$, and that this fall-off is smooth except for Fraunhofer-type diffractive oscillations which are, however, normally strongly damped by the function $b(\theta)$.

Taking as a specific example the form (6.3.17) of $F_N(\Delta z)$, we have

$$\left|\frac{F_N\{\Delta(\theta_R - \theta)\}}{\theta_R - \theta}\right|^2 = (\pi\Delta)^2 \frac{\exp\{-2\alpha\Delta(\theta - \theta_R)\}}{[\sinh\{\pi\Delta(\theta - \theta_R)\}]^2} \tag{6.5.14}$$

and

$$b(\theta) = -\frac{\sinh\{\pi\Delta(\theta-\theta_R)\}}{\sinh\{\pi\Delta(\theta+\theta_R)\}}\exp(2\alpha\Delta\theta). \qquad (6.5.15)$$

Then, if Δ is not too small, the cross-section becomes

$$\sigma(\theta) \simeq \frac{2\pi\Delta^2\Lambda}{k^2\sin\theta}\exp\{-2(\pi+\alpha)\Delta(\theta-\theta_R)\}\{1+\exp(-4\pi\Delta\theta_R)\exp(4\alpha\Delta\theta)$$
$$-2\exp(-2\pi\Delta\theta_R)\exp(2\alpha\Delta\theta)\sin(2\Lambda\theta)\}. \qquad (6.5.16)$$

This result displays several significant features. First, the slope of the exponential decrease of $\sigma(\theta)$ plotted as a function of $\theta-\theta_R$ is given by $2(\pi+\alpha)\Delta$, and thus is proportional to the width Δ of the l-window, and is made steeper by the nuclear phase parameter $\alpha>0$. Further, the amplitude of the Fraunhofer oscillations is strongly damped by the Coulomb field, as shown by the exponential $2\pi\Delta\theta_R = 4\pi\Delta$ arctan (n/Λ), but this damping is counteracted by the nuclear interaction, similar to what is found for the Fresnel oscillations at $\Theta<\theta_R$.

6.5.3 *Near the grazing angle θ_R: the quarter-point property*

We now investigate more closely the behaviour of eqns (6.5.1) in the neighbourhood of θ_R. With the definitions (6.4.33), (6.4.34), (6.5.2), and (6.5.7) we obtain

$$\frac{f^{(+)}(\theta)}{f_R(\theta)} = 1-\tfrac{1}{2}\operatorname{erfc}\{-\exp(\mathrm{i}\tfrac{1}{4}\pi)u\}c(\theta)\{1-(\theta-\theta_R)a_1(u)\}\exp\{-\mathrm{i}\psi(u)\}$$
$$+A(\theta)\left[\frac{F_N\{\Delta(\theta_R-\theta)\}-1}{\theta_R-\theta}-a_0(u)\right], \qquad \theta\leqslant\theta_R, \qquad (6.5.17a)$$

$$\frac{f^{(+)}(\theta)}{f_R(\theta)} = \tfrac{1}{2}\operatorname{erfc}\{\exp(\mathrm{i}\tfrac{1}{4}\pi)u\}c(\theta)[1-(\theta-\theta_R)a_1(u)]\exp\{-\mathrm{i}\psi(u)\}$$
$$+A(\theta)\left[\frac{F_N\{\Delta(\theta_R-\theta)\}-1}{\theta_R-\theta}-a_0(u)\right] \qquad \theta\geqslant\theta_R. \qquad (6.5.17b)$$

By using the property of the error function

$$\tfrac{1}{2}\{\operatorname{erfc}(z)+\operatorname{erfc}(-z)\}=1 \qquad (6.5.18)$$

it is seen that the right-hand sides of (6.5.17*a*) and (6.5.17*b*) differ by

$$1-c(\theta)\{1-(\theta-\theta_R)a_1(u)\}\exp\{-\mathrm{i}\psi(u)\} = O((\theta-\theta_R)^2), \qquad (6.5.19)$$

which shows that our expressions are differentiably continuous at $\theta=\theta_R$.

For the value at the grazing angle we find

$$\frac{f(\theta_R)}{f_R(\theta_R)}=\frac{1}{2}+A(\theta_R)\left\{\Delta F_N'(0)-a_0(0)-\mathrm{i}\frac{F_N(2\Delta\theta_R)}{2\theta_R}\exp(\mathrm{i}2\Lambda\theta_R)\right\}$$

$$=\frac{1}{2}\left[1+\frac{\exp(-\mathrm{i}\frac{1}{4}\pi)}{2(\pi n)^{1/2}}\frac{1}{\cos\frac{1}{2}\theta_R}\left\{1+\tfrac{4}{3}(\cos\tfrac{1}{2}\theta_R)^2-2\Delta F_N'(0)\sin\theta_R\right.\right.$$

$$\left.\left.+\mathrm{i}F_N(2\Delta\theta_R)\frac{\sin\theta_R}{\theta_R}\exp(\mathrm{i}2\Lambda\theta_R)\right\}\right]. \tag{6.5.20}$$

This result expresses in analytic form the 'quarter-point property' of the strong-absorption scattering of charged particles, i.e. $\sigma/\sigma_R\approx\frac{1}{4}$ at $\theta=\theta_R$, which was first found empirically by Blair (1954 [6.10]). However, our expression shows that this property holds only up to corrections of order $n^{-1/2}$; these corrections should be taken into account in practical applications of the quarter-point recipe for the determination of the critical angular momentum Λ from experimental data.

On expanding eqns (6.5.17*a*), (6.5.17*b*), and (6.5.1*c*) about $\theta=\theta_R$, the term of leading order in Λ or n arises from $\mathrm{erfc}(z)=1-(2/\pi^{1/2})z+O(z^3)$, and we obtain in the linear neighbourhood of θ_R

$$\frac{f(\theta)}{f_R(\theta)}\approx\frac{f(\theta_R)}{f_R(\theta_R)}-\left(\frac{n}{\pi}\right)^{1/2}\frac{\exp(\mathrm{i}\frac{1}{4}\pi)}{2\sin\frac{1}{2}\theta}(\theta-\theta_R). \tag{6.5.21}$$

Abbreviating $\sigma(\theta)/\sigma_R(\theta)$ by $\rho(\theta)$ we can write

$$\rho(\theta)\approx\rho(\theta_R)\left\{1-\left(\frac{2n}{\pi}\right)^{1/2}\frac{\theta-\theta_R}{\sin\frac{1}{2}\theta_R}\right\}. \tag{6.5.22}$$

If we denote the true quarter-point angle by $\theta_{1/4}$, so that $\rho(\theta_{1/4})=\frac{1}{4}$, we find from eqns (6.5.22) and (6.5.20),

$$\theta_{1/4}-\theta_R\approx\left(\frac{\pi}{2n}\right)^{1/2}\sin\tfrac{1}{2}\theta_R\left\{1-\frac{1}{4\rho(\theta_R)}\right\}$$

$$\approx\frac{1}{2n}\tan\tfrac{1}{2}\theta_R\left\{1+\tfrac{4}{3}(\cos\tfrac{1}{2}\theta_R)^2-2\Delta F_N'(0)\sin\theta_R\right.$$

$$\left.+\sqrt{2}F_N(2\Delta\theta_R)\frac{\sin\theta_R}{\theta_R}\cos(2\Lambda\theta_R+\tfrac{1}{4}\pi)\right\}. \tag{6.5.23}$$

6.5.4 $\mathcal{N}$-limit: Fraunhofer-diffractive regime

This limit is approached with increasing energy E, since n/Λ varies roughly as E^{-1}. As θ_R decreases, the 'shadow region' extends towards smaller angles, and eqns (6.5.12) and (6.5.13) hold for most of the angular distribution. However, because the Coulomb damping weakens, the far-side branch of $f(\theta)$ increases in relative magnitude and the Fraunhofer oscillations become more and more pronounced. At the same time the nuclear phase contribution begins to change its role: while in the $\mathcal{C}$-limit it counteracts the effect of Coulomb damping and enhances the oscillatory part of the cross-section, in the $\mathcal{N}$-limit it tends to diminish the Fraunhofer oscillations by 'refractive damping'.

The see this change clearly, we write the cross-section in a form in which the contributions of the near-side and far-side branches are treated on the same footing, and by using the asymptotic expression of our model (6.3.17) for $F_N(\Delta z)$. This reads

$$\sigma(\theta) \simeq \frac{2\pi\Delta^2\Lambda}{k^2 \sin\theta} \exp\{-2\pi\Delta(\theta-\theta_R)\}\exp\{2\alpha\Delta(\theta+\theta_R)\} \\ \times |\exp(-2\alpha\Delta\theta)\exp(-i\Lambda\theta) + i\exp(-2\pi\Delta\theta_R)\exp(i\Lambda\theta)|^2 \quad (6.5.24)$$

where the first term in the modulus comes from the (+) branch and the second term from the (−) branch of the scattering amplitude. Thus the Coulomb field damps the far-side amplitude and favours the near-side branch, whereas the nuclear field has the opposite effect. In the extreme $\mathcal{N}$-limit $\theta_R = 0$, eqn (6.5.24) reduces to

$$\sigma(\theta) \simeq \frac{2\pi\Delta^2\Lambda}{k^2 \sin\theta} \exp\{-2(\pi-\alpha)\Delta\theta\} \\ \times\{1 + \exp(-4\alpha\Delta\theta) - 2\exp(-2\alpha\Delta\theta)\sin(2\Lambda\theta)\}, \quad (6.5.25)$$

which displays the effect of 'refractive damping' of the diffraction oscillations.

For general $F_N(\Delta z)$ the amplitude (6.5.12) in the shadow region can be written in the form

$$f(\theta) = \frac{\kappa(\theta)}{k}\left[\frac{F_N\{\Delta(\theta_R-\theta)\}}{\theta_R-\theta}\exp\{i\phi_+(\Lambda,\theta)\} + \frac{F_N\{\Delta(\theta_R+\theta)\}}{\theta_R+\theta}\exp\{i\phi_-(\Lambda,\theta)\}\right]. \quad (6.5.26)$$

In the extreme $\mathcal{N}$-limit this becomes

$$f(\theta) = \left(\frac{\Lambda}{2\pi\sin\theta}\right)^{1/2} \frac{1}{k\theta}[-F_N(-\Delta\theta)\exp\{-i(\Lambda\theta - \tfrac{1}{4}\pi)\} \\ + F_N(\Delta\theta)\exp\{i(\Lambda\theta-\tfrac{1}{4}\pi)\}], \quad (6.5.27)$$

in agreement with what is obtained from eqn (6.3.12) for the 'neutral' scattering amplitude if only the $m = 0$ terms in the Poisson series (6.3.13) are retained and the asymptotic expressions (6.3.14) are used, consistent with the assumptions underlying our derivation in Section 6.4.

6.5.5 Pictorial explanation

The analytic results of the previous subsections are illustrated in Figs. 6.5–6.7. Figure 6.5 shows, on a linear scale, the cross-section ratio σ/σ_R in the illuminated region; in order to display most clearly the Fresnel oscillations and their modulation by the smaller and more rapid Fraunhofer oscillations, the curves are calculated with $\Delta = 0$, representing the 'sharp-cut-off' limit. For finite Δ these oscillations are damped, the Fresnel ones by the function $F_N\{\Delta(\theta_R - \theta)\}$, and the Fraunhofer ones much more strongly by the function $F_N\{\Delta(\theta_R + \theta)\}$. Figure 6.6 shows σ/σ_R on a logarithmic scale, for finite Δ, and depicts the strong smooth fall-off in the shadow region $\theta > \theta_R$. Also indicated is the small difference between the grazing angle θ_R and the quarter-point angle $\theta_{1/4}$. To display the effects of the nuclear phase parameter α, the angular distributions in Fig. 6.7 are calculated with a value of Δ small enough to bring out the Fraunhofer oscillations in the shadow region; it is shown how $\alpha > 0$

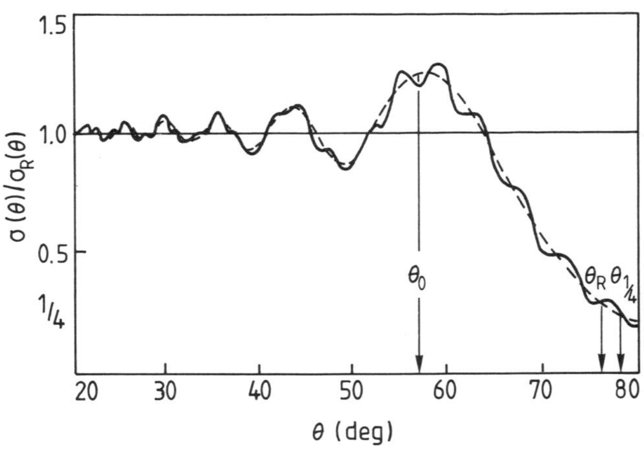

FIG. 6.5. Differential cross section ratio $\sigma(\theta)/\sigma_R(\theta)$ in the 'illuminated region' $\theta < \theta_R$ calculated from the closed formulae in the sharp-cut-off limit $\Delta = 0$ (eqns (6.4.32)): ---, the Fresnel diffraction pattern obtained by averaging over the Fraunhofer-type oscillations. The location of the Rutherford grazing angle θ_R, the quarter-point angle $\theta_{1/4}$, and the maximum θ_0 of the first Fresnel peak are indicated.

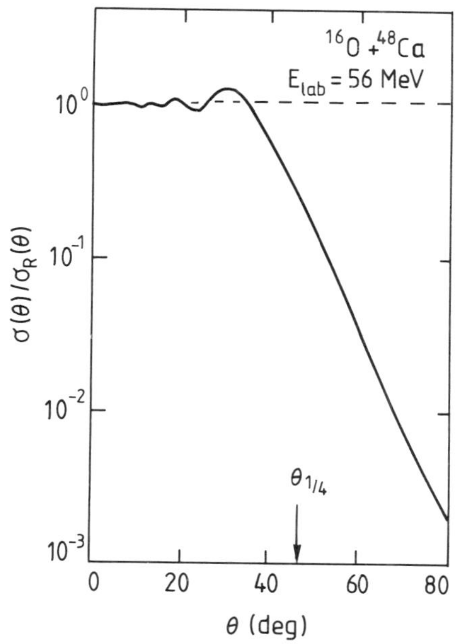

FIG. 6.6. Differential cross-section ratio $\sigma(\theta)/\sigma_R(\theta)$ calculated from the closed formulae (eqns (6.5.1)). The figure shows the rapid smooth fall-off of the angular distribution in the shadow region $\theta > \theta_R$ under conditions of strong Coulomb damping.

steepens the slope of σ/σ_R and at the same time enhances the amplitude of the oscillations with increasing θ. The opposite effect of α in the $\mathcal{N}$-limit has already been shown in Fig. 6.4.

Our closed-form expressions suggest simple qualitative explanations of these features. Firstly, the reason why Fresnel diffraction occurs at all in strong-absorption scattering of charged particles has already been explained in Chapter 1 by the 'diverging-lens' effect of the Coulomb field surrounding the nuclear surface. The difference in the structure of the angular distribution in the illuminated and shadow regions can be described as follows.

Consider first the sharp-cut-off limit. We have seen that in the region $\theta < \theta_R$ there are two main contributions to the scattering amplitude: one from the point of stationary phase, $\lambda_\theta = n \cot \frac{1}{2}\theta$, and the other from the vicinity of the grazing angular momentum $\Lambda = n \cot \frac{1}{2}\theta_R$. The former contribution can be represented by a classical *trajectory* of impact parameter $b_\theta = \lambda_\theta/k$, and the latter by a *diffracted wave* originating from the point at which the grazing trajectory of impact parameter $B = \Lambda/k$ touches the nuclear surface. This diffracted wave is centred about the

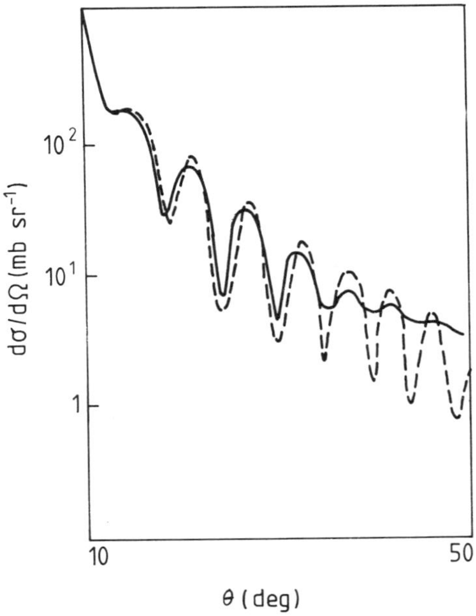

FIG. 6.7. Differential scattering cross-section $\sigma(\theta)$ calculated from the closed formulae (6.4.39) for $\alpha = 0$ and $\alpha > 0$. The figure illustrates the effects of the nuclear phase parameters in the intermediate regime between Fresnel and Fraunhofer scattering, and should be compared with Fig. (6.4) which shows the effects of α in the extreme Fraunhofer diffractive region.

direction θ_R, rather than about the forward direction $\theta = 0$ as for neutral particles, because of the outward bending of the diffracted rays by the Coulomb field. This situation is sketched in Fig. 6.8. The interference between the contributions from the trajectory in the variable direction θ and from the 'fan' of diffracted rays around the fixed direction θ_R produces the Fresnel diffraction pattern in the region $\theta < \theta_R$. In addition there is a small oscillatory contribution of Fraunhofer type arising from the interference of the two diffracted waves originating at the near side and at the far side of the interaction region.

In the shadow region $\theta > \theta_R$ there is no contribution from a stationary point (the corresponding trajectories have impact parameters $b_\theta < R$ and are completely 'absorbed'), so the only contributions are from the near-side and far-side diffracted waves. However, the degree of interference between these amplitudes depends on the extent of overlap between the two diffraction 'fans', and this decreases with increasing θ_R. As a result, the Fraunhofer oscillations show some 'Coulomb damping' even in the sharp-cut-off limit.

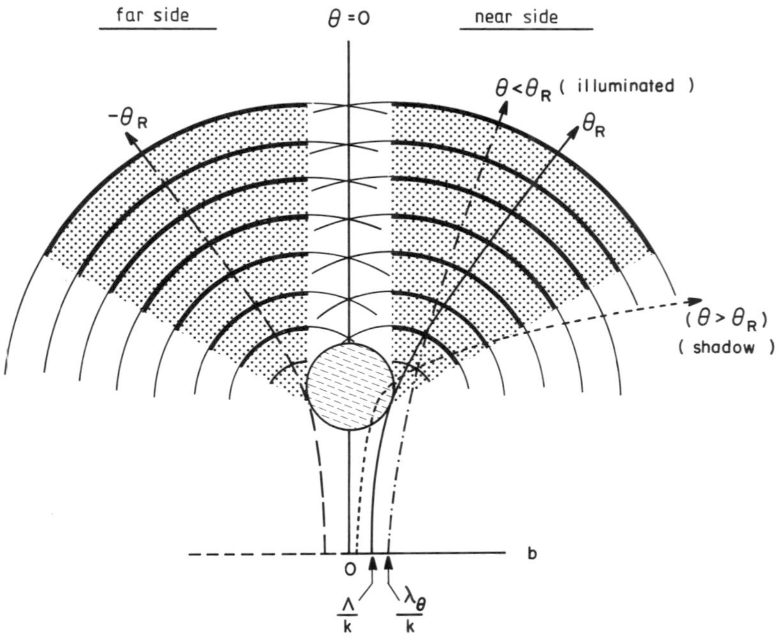

FIG. 6.8. Origin of quantal interference and diffraction patterns in the strong-absorption scattering of charged particles (sharp-cut-off limit). In the 'illuminated region' $\theta < \theta_R$ the Rutherford trajectory scattering into θ (of impact parameter $b_\theta = \lambda_\theta/k$, indicated by the chain curve) interferes with the near-side diffracted wave originating from the point where the grazing trajectory (of impact parameter $B = \Lambda/k$) touches the nuclear surface, giving rise to a Fresnel diffraction pattern. This pattern is modulated by Fraunhofer oscillations arising from interference between the near-side and far-side diffracted waves (the light circles represent the wave fronts of the diffracted waves). In the shadow region $\theta > \theta_R$ only the Fraunhofer pattern occurs, since possible trajectories of impact parameter $b < B$ passing through the nuclear interior (indicated by the dotted curve) are completely absorbed.

Now let us take account of the finite 'window' of width Δ in l-space, to which corresponds a width $d = \Delta/k$ in impact parameter space. Its effect on the diffracted waves is determined by the uncertainty relation: the wider is Δ, the narrower is the diffraction pattern. Note the significant point that this reciprocity between the l-space profile and the θ-space pattern is embodied in our formalism by the Fourier transforms $F_N\{\Delta(\theta_R \mp \theta)\}$, and that this is the reason why the details of the nuclear interaction enter only through these functions! Thus, with increasing Δ the fans of diffracted rays contract and their overlap diminishes. In the

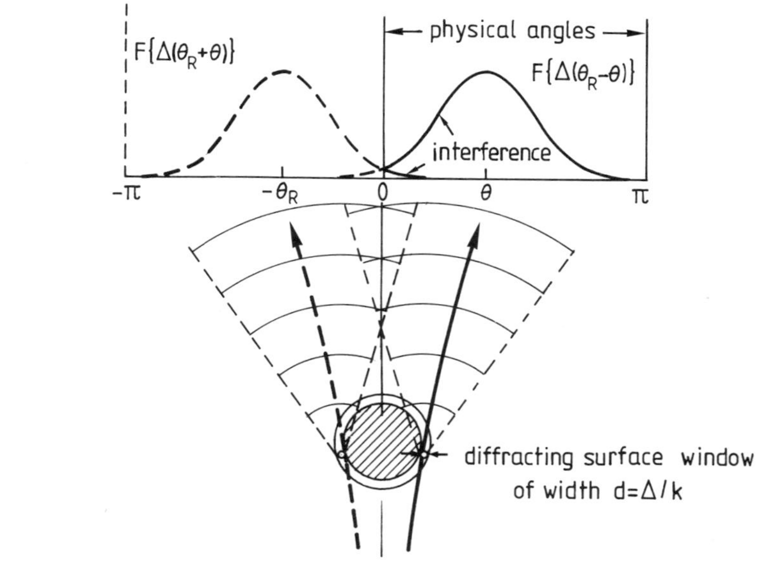
physical angles
F{Δ(θR+θ)}
F{Δ(θR−θ)}
interference
−π
−θR
0
θ
π
diffracting surface window
of width d=Δ/k

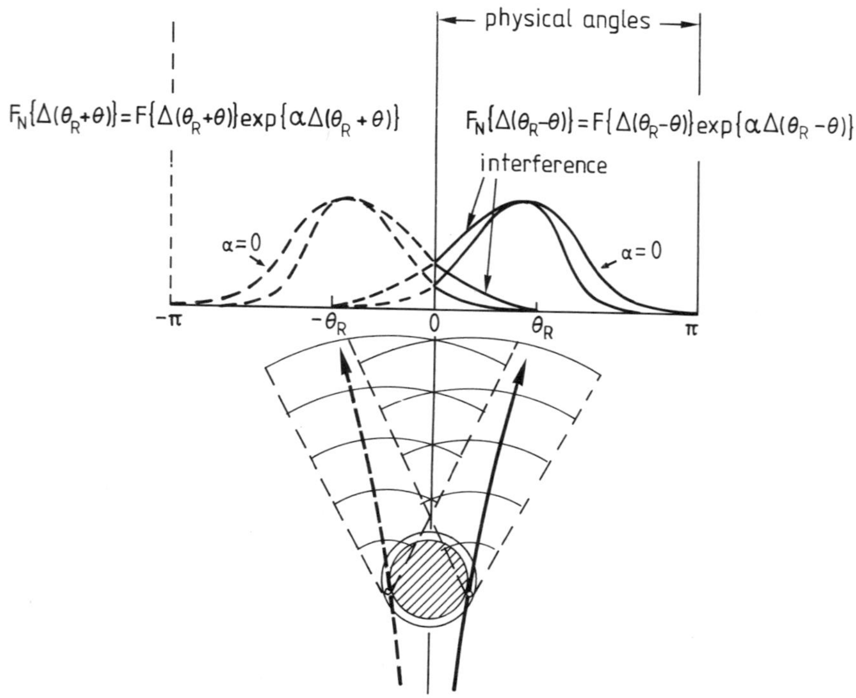
physical angles
FN{Δ(θR+θ)}=F{Δ(θR+θ)}exp{αΔ(θR+θ)}
FN{Δ(θR−θ)}=F{Δ(θR−θ)}exp{αΔ(θR−θ)}
interference
α=0
α=0
−π
−θR
0
θR
π

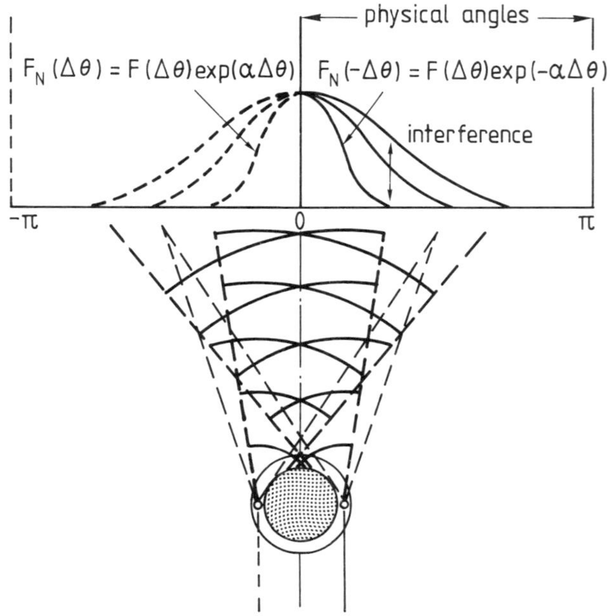

FIG. 6.9. (a) Diffraction scattering of charged particles through an l-space window of finite width Δ. The near-side and far-side diffracted waves centred about the directions θ_R and $-\theta_R$ are contracted, with a profile described by the Fourier transform $F(\Delta Z)$ because of the uncertainty relation between angular momentum and angle. This contraction results in a damping of the Fraunhofer diffraction pattern in the region of physical angles ('Coulomb damping'). (b) Refractive enhancement of the Fraunhofer oscillations in charged particle scattering. The effect of the nuclear phase parameter $\alpha > 0$ is to render the profile of the diffracted waves asymmetrical about θ_R and $-\theta_R$ through an exponential factor in the Fourier transform $F_N(\Delta Z) = F(\Delta Z)\theta^{\alpha\,\Delta Z}$. Because of the increased overlap between the near-side and far-side diffraction profiles in the region of physical angles the Fraunhofer oscillations are enhanced. (c) 'Refraction damping' of the Fraunhofer oscillations in neutral-particle scattering. While for $\alpha = 0$ the profiles $F(\Delta Z)$ of the near-side and far-side diffracted waves are symmetrical about $\theta = 0$ (or nearly so), and thus have maximal overlap, a positive value of the phase parameter renders them asymmetrical, suppressing the near-side profile and enhancing the far-side profile. Because of the reduced overlap the Fraunhofer pattern becomes increasingly damped with increasing angle.

region $\theta < \theta_R$ this results in the damping of the Fresnel pattern as well as in the even stronger damping of the superimposed Fraunhofer oscillations. In the region $\theta > \theta_R$ the overlap between the near-side and far-side diffraction fans is greatly reduced; this is the reason for the 'Coulomb damping' of the Fraunhofer pattern and explains why it is controlled by the product $\Delta\theta_R$. In fact, normally the contraction is so strong that the overlap is imperceptible, which accounts for the smooth fall-off of the cross-section in the shadow region.

The effects of the nuclear phases are readily explained by their origin from an attractive nuclear field. This causes the bundles of diffracted rays to be bent inwards, opposing the outward bending by the repulsive nuclear field. This increases the overlap between the near-side and far-side diffraction fans and enhances the Fraunhofer oscillations. At the same time it sharpens the angular pattern and increases the rate of fall-off of the average cross-section in the shadow region. However, the nuclear phase effects can operate in this fashion only if the Coulomb interaction is appreciable. In the neutral or near-neutral case, where the overlap between the near-side and far-side diffraction fans is already maximal—in the forward direction—the effect of 'inward bending' by the nuclear field is to diminish the overlap. This explains the change from 'Coulomb damping' of the diffraction oscillations in the $\mathscr{C}$-limit to the 'refractive damping' in the $\mathscr{N}$-limit. All these effects are illustrated in Fig. 6.9.

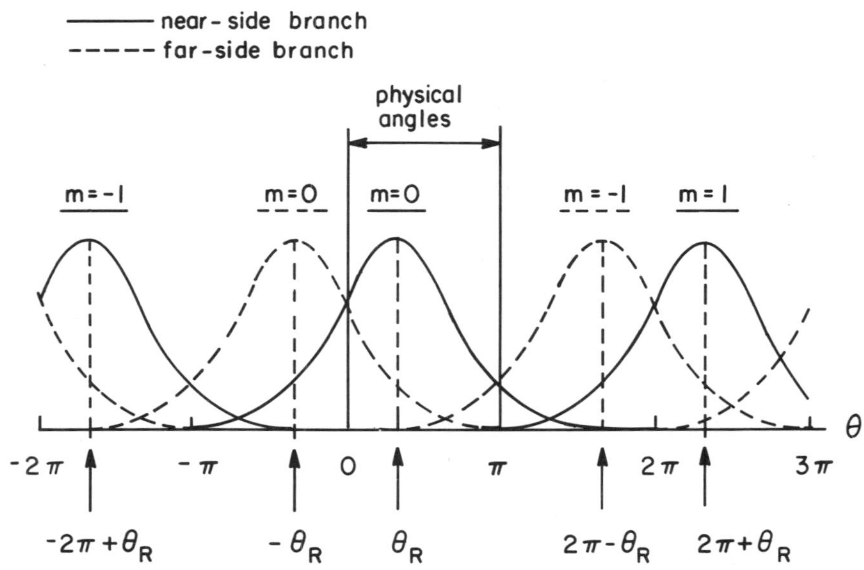

FIG. 6.10. Profiles of the diffracted waves corresponding to different terms in the Poisson series for the elastic scattering amplitude for charged particles. These are given by the Fourier transforms $F_N\{\Delta(\theta_R+2m\pi-\theta)\}$ in the near-side (+) branch (——) and $F_N\{\Delta(\theta_R+2m\pi+\theta)\}$ in the far-side (−) branch (– – –), (For simplicity a symmetrical profile is shown.) It is evident that only the $m=0$ terms are significant, except for large-angle scattering where there is a contribution from the $m=-1$ term in the (−) branch. All other terms have negligible overlap in the region of physical angles.

Finally, an inspection of Fig. 6.10, in which different contributions to the Poisson series are sketched, shows why, under strong absorption conditions, only the $m=0$ terms contribute significantly to 'forward' scattering, and both the $m=0$ and $m=-1$ terms contribute to backward scattering.

The qualitative discussion given here is just a verbal description of the analytic features of the formulae derived in the previous sections. Conversely, the structure of our closed-form expressions suggests unequivocally our physical interpretation, and there can be little doubt about its validity. Of course, our considerations so far apply only to 'normal strong-absorption scattering', i.e. scattering in the absence of special dynamical effects which will be treated in Chapters 8 and 9.

6.6. Spin polarization

So far in this chapter we have either considered zero-spin nuclei only or disregarded spin effects. In this section we deal with the spin dependence of the heavy-ion interaction, confining ourselves for simplicity to the scattering of projectiles of general spin s by zero-spin targets and to (first-order) spin–orbit coupling.

6.6.1 *Early predictions*

A first attempt to describe spin effects in heavy-ion scattering was made in 1963 by an extension of the SAM formalism (Frahn and Venter 1964 [6.11]). For spin-$\frac{1}{2}$ projectiles a spin dependence was introduced into the S-matrix by assuming that the parameters Λ, Δ, μ, etc. in the scattering function (6.2.7) have different values in the two states $j=l+\frac{1}{2}$ and $j=l-\frac{1}{2}$ of angular momentum $\mathbf{j}=\mathbf{l}+\mathbf{s}$, but that otherwise the forms of $\mathrm{Re}\, S_N(\lambda)$ and $\mathrm{Im}\, S_N(\lambda)$ are the same as for spin-zero nuclei. Closed-form expressions were derived for the spin-flip and spin-nonflip amplitudes and hence for the vector polarization and differential cross-section. If it is assumed that the spin–orbit interaction is confined to the nuclear surface and is predominantly real, its main effect will be to produce differences in the strength of the effective potential in the surface and in the interaction radii, and thus in the phase parameter μ and the grazing angular momentum in the states $j=l\pm\frac{1}{2}$. Accordingly, the expressions for the polarization $P(\theta)$ contained terms proportional to the parameter differences $\nu=\mu_+-\mu_-$ and $\tau=\Lambda_+-\Lambda_-$, whose values are measures of the strength of the spin–orbit coupling.

The only heavy projectile with non-zero spin available at the time was ^{19}F ($s=\frac{1}{2}$), and predictions were made for the polarization in the system $^{19}\mathrm{F}+{}^{159}\mathrm{Tb}$ at $E_{\mathrm{lab}}=159$ MeV for which an elastic angular distribution

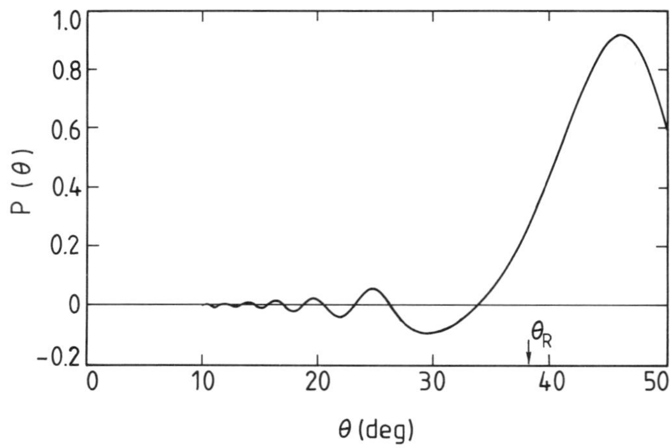

FIG. 6.11. Early prediction of the angle dependence of the spin polarization $P(\theta)$ in heavy-ion scattering, showing strong destructive Coulomb nuclear interference in the Fresnel region $\theta < \theta_R$. The curve was calculated using the closed expressions of the strong-absorption model for spin-$\frac{1}{2}$ particles for the scattering of $^{19}F + ^{159}Tb$ at $E_{lab} = 159$ MeV. (From Frahn and Venter 1964 [6.11].)

had been measured (McIntyre *et al.* 1962 [6.12]) with the (tentative) assumption that the spin–orbit interaction for heavy-ion nucleus scattering has the same strength as for nucleon–nucleus scattering. The result (Fig. 6.11) shows that the polarization is very small at small angles and remains so, with some (Fresnel-type) oscillations, up to the grazing angle θ_R, but attains high values in the shadow region well beyond θ_R. However, since the cross-section in this region has already dropped two orders of magnitude below the Rutherford values, a measurement of the polarization seemed hopeless in a double-scattering experiment even if polarized ion sources had been available.

6.6.2 *Generalized closed-form description*

In 1975 Fick and his coworkers at the Heidelberg EN tandem accelerator (Weiss *et al.* 1976 [6.13]) were able to measure vector polarization in heavy-ion scattering for the first time. They used a beam of polarized ^6Li ions with $E_{lab} = 22.8$ MeV. The qualitative predictions of Frahn and Venter (1964 [6.11]) were confirmed for the heavier target nuclei (^{28}Si and ^{50}Ni). Optical model analyses with folded (Weiss *et al.* 1976 [6.13]; Petrovich *et al.* 1978 [6.14]) and conventional (Meyer *et al.* 1978 [6.15]) potentials gave reasonable fits to the data but were somewhat inconclusive with regard to the strength and form of the spin–orbit potential.

Although a simple interpretation of the polarization as a shape effect (Fick 1976 [6.16]), based on Fresnel model considerations, gave some qualitative insight, it was clear that an adequate S-matrix description was called for to account for the Fraunhofer-diffractive features observed for the lighter target nuclei. Since ^{6}Li has spin $s=1$, a generalization of the SAM formulation was required. An extension of the closed formalism described in Section 6.2 to include the spin–orbit interaction of projectiles of any spin s in the scattering function was developed by Hill and Frahn (1980 [6.17]); the relevant results of this work and their physical interpretation will be discussed here.

In a coordinate system with the quantization axis normal to the scattering plane (i.e. the x axis in direction $\mathbf{k}$ and the z axis in direction $\mathbf{k}\times\mathbf{k}'$), the S-matrix S_{jl} for spin-s projectiles and spin-zero targets can be replaced by the scattering functions $S_\nu(\lambda)$:

$$S_{jl}=S_{l+\nu,l}\rightarrow S_\nu(\lambda), \tag{6.6.1}$$

where ν denotes the z component of the spin $\mathbf{s}$. Assuming that the spin dependence can be factored out, we can write

$$S_\nu(\lambda)=S_c(\lambda)\exp\{2i\nu\delta_s(\lambda)\} \tag{6.6.2}$$

where $S_c(\lambda)$ is the scattering function for the central interaction, and the *complex* phase shift function $\delta_s(\lambda)$ arises from a complex spin–orbit interaction. For simplicity we confine ourselves to first-order spin–orbit coupling by the condition

$$|2s\delta_s(\lambda)|\ll 1. \tag{6.6.3}$$

Then the differential scattering cross-section becomes

$$\sigma(\theta)=|f_0(\theta)|^2+\tfrac{1}{3}s(s+1)\,|f_1(\theta)|^2=\sigma_0(\theta)+\sigma_1(\theta) \tag{6.6.4}$$

and the vector analysing power $A(\theta)=P(\theta)$ is given by

$$A(\theta)=\frac{1}{\sigma(\theta)}\frac{2}{3}(s+1)\mathrm{Re}\{f_0(\theta)f_1^*(\theta)\} \tag{6.6.5}$$

where $f_0(\theta)$ is the spin-independent amplitude and $f_1(\theta)$ is the first-order spin-dependent amplitude.

$f_0(\theta)$ is given by the closed expressions (6.4.39), and the amplitude $f_1(\theta)$ is calculated as

$$f_1(\theta)=-\frac{2\tilde{\kappa}(\theta)}{k}[\mathscr{P}(\tilde{\theta}_R-\theta)\exp\{i\phi_+(\tilde{\Lambda},\theta)\}-\mathscr{P}(\tilde{\theta}_R+\theta)\exp\{-i\phi_-(\tilde{\Lambda},\theta)\}] \tag{6.6.6}$$

where $\mathcal{P}(x)$ is the Fourier transform of the function $S_{N,c}(\lambda)\delta_s(\lambda)$,

$$\mathcal{P}(x)=\int_{-\infty}^{\infty} d\lambda S_{N,c}(\lambda)\delta_s(\lambda)\exp\{i(\lambda-\tilde{\Lambda})x\}, \tag{6.6.7}$$

and

$$\tilde{\kappa}(\theta)=\left(\frac{\tilde{\Lambda}}{2\bar{\kappa}\sin\theta}\right)^{1/2}, \qquad \tilde{\theta}_R=2\arctan\left(\frac{n}{\tilde{\Lambda}}\right) \tag{6.6.8}$$

$$\phi_{\pm}(\tilde{\Lambda},\theta)=2\sigma(\tilde{\Lambda})\mp(\tilde{\Lambda}\theta-\tfrac{1}{4}\pi) \tag{6.6.9}$$

with $\tilde{\Lambda}$ defined as the location of the maximum of $|S_{N,c}(\lambda)\delta_s(\lambda)|$. Because the spin–orbit interaction is confined to the nuclear surface, the latter function is localized in the 'surface' region of Λ-space; thus $\tilde{\Lambda}$ is expected to be close to the grazing parameter Λ, and $\tilde{\theta}_R$ will not be far from the grazing angle θ_R.

For practical calculations we adopt the Ericson parametric form (6.2.7) for $S_{N,c}(\lambda)$, and for the spin–orbit phase function we assume

$$2\delta_s(\lambda)=4\kappa\Delta_s\frac{d}{d\lambda}\left[\left\{1+\exp\left(\frac{\Lambda_s-\lambda}{\Delta_s}\right)\right\}^{-1}\right] \tag{6.6.10}$$

where the (complex) constant κ measures the strength of the spin–orbit coupling. The form (6.6.10) is motivated by the high-energy (eikonal) approximation, if the spin–orbit interaction has the customary form $U_{so}(r)\mathbf{l}\cdot\mathbf{s}$, with

$$U_{so}(r)=U_s\left(\frac{\hbar}{m_\pi c}\right)^2\frac{1}{r}\frac{df_s(r)}{dr}$$

$$f_s(r)=\left\{1+\exp\left(\frac{r-R_s}{a_s}\right)\right\}^{-1}. \tag{6.6.11}$$

Then the strength constant κ is approximately related to U_s by

$$\kappa=\kappa_R+i\kappa_I=\frac{\alpha}{\Delta_s}\left(\frac{\hbar k}{2m_\pi c}\right)^2\frac{U_s}{V}, \tag{6.6.12}$$

where α is the phase parameter in $S_{N,c}(\lambda)$ and V is the depth of the real part of the central potential. The condition (6.6.3) implies $|\kappa|\ll s^{-1}$. Analytic expressions for the function $\mathcal{P}(x)$ are given by Hill and Frahn (1980 [6.17, 6.18]).

The spin-dependent amplitude (6.6.6) has a simple structure consisting of a near-side and a far-side branch whose magnitudes and angular dependences are determined by the Fourier transforms $\mathcal{P}(\tilde{\theta}_R-\theta)$ and $\mathcal{P}(\tilde{\theta}_R+\theta)$ respectively. From eqn (6.6.4) we obtain for the spin–orbit contribution to the cross-section

$$\sigma_1(\theta)=\frac{4}{3}s(s+1)\frac{\{\tilde{\kappa}(\theta)\}^2}{k^2}|\mathscr{P}(\tilde{\theta}_R-\theta)+i\mathscr{P}(\tilde{\theta}_R+\theta)\exp(i2\tilde{\Lambda}\theta)|^2. \tag{6.6.13}$$

Since this expression is of second order in κ, it is a small correction to $\sigma_0(\theta)$ at small and intermediate angles. However, in the *large-angle region* it may be quite appreciable: since $\mathscr{P}(x)$ is the Fourier transform of a relatively sharply localized function in λ-space, it describes a broad distribution in θ which may dominate over the strongly decreased $\sigma_0(\theta)$ at large angles. Moreover, eqn (6.6.13) shows that there are regular Fraunhofer-type diffraction oscillations of period $\pi/\tilde{\Lambda}$ which, again because of the sharp λ localization, will not be strongly damped. *These predictions are worth testing experimentally.*

For the vector analysing power $A(\theta)$ we obtain, from eqns (6.6.5), (6.6.6) and (6.6.8), with some simplifications,

$$A(\theta)\approx\frac{1}{\sigma(\theta)k^2}\frac{4}{3}s(s+1)\frac{\Lambda^{3/2}}{(2\pi\sin\theta)^{1/2}}\frac{1}{\sin\theta}\operatorname{Re}\left[\mathscr{P}^*(\tilde{\theta}_R-\theta)\exp\left\{i\left(u^2+\frac{\pi}{4}\right)\right\}\right] \tag{6.6.14}$$

in the region $\theta<\theta_R$. This describes Fresnel-type oscillations with slowly growing amplitude about a very small mean value. On the other hand, in the shadow region $\theta>\theta_R$ we have approximately

$$A(\theta)\approx-\frac{1}{\sigma(\theta)k^2}\frac{4}{3}d(s+1)\kappa(\theta)\tilde{\kappa}(\theta)$$
$$\times\operatorname{Re}[\{G_+\exp(i\phi_++G_-\exp(i\phi)\}\{\mathscr{P}_+(\exp(i\tilde{\phi}_+)-\mathscr{P}_-\exp(i\tilde{\phi}_-)\}^*] \tag{6.6.15}$$

where we have used the abbreviations $G_\pm=G_N(\theta_R\mp\theta)$, $P_\pm=P(\tilde{\theta}_R\mp\theta)$, $\phi_\pm=\phi_\pm(\Lambda,\theta)$ and $\tilde{\phi}_\pm=\phi_\pm(\tilde{\Lambda},\theta)$. This shows that $A(\theta)$ has a Fraunhofer-type oscillatory structure with approximate period π/Λ, for $\theta>\theta_R$, in the Fresnel as well as in the Fraunhofer scattering regime.

More details concerning spin polarization in elastic scattering and the corresponding data analysis are given in the paper by Hill and Frahn (1980 [6.17])

6.6.3 *Data analysis and interpretation*

An application of this formalism to the data for $^6\vec{Li}+^{28}Si$ and $^6Li+^{58}Ni$ is shown in Figs. 6.12 and 6.13 with the parameter values given in Table 6.6.1. The features of $A(\theta)$ and their relation to those of σ/σ_R described by the formulae, in particular the presence of both Fresnel and

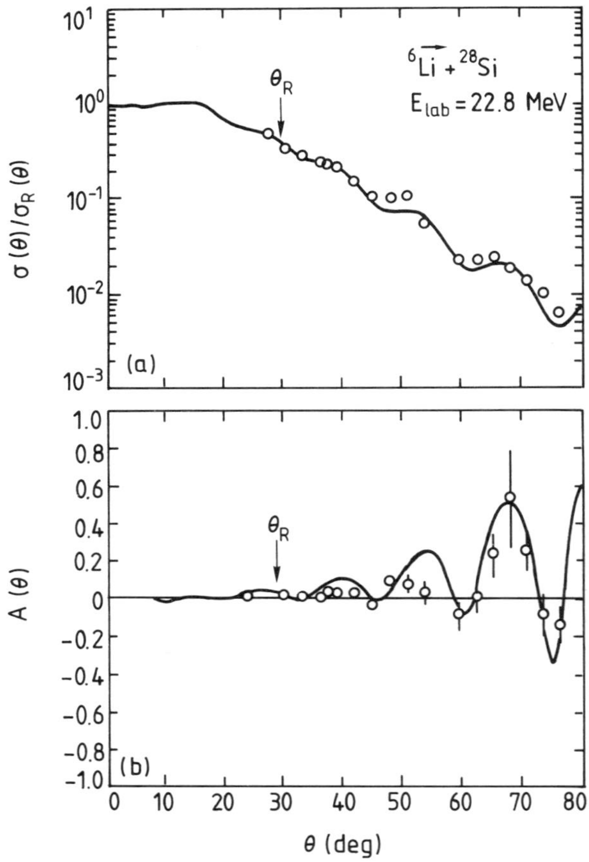

FIG. 6.12. (a) Differential cross-section ratio and (b) vector analysing power for the scattering of $^6\vec{\mathrm{Li}}+{}^{28}\mathrm{Si}$ at $E_{\mathrm{lab}}=22.8$ MeV. The data are from Weiss *et al.* (1976 [6.13]), and the curves are calculated from the closed expressions given in Section 6.6.2 using the parameter values listed in Table 6.6.1. (From Hill and Frahn (1980 [6.17]).)

Fraunhofer diffractive structure, are clearly borne out. The values of κ in Table 6.6.1 indicate the presence of a significant imaginary part of the spin–orbit interaction, which is not surprising in strong-absorption scattering. Further, comparison with similar analyses of $^3\vec{\mathrm{He}}$ scattering suggests that the strength of the spin–orbit coupling for heavy ions is not weaker than that for light ions, contrary to the predictions of simple folding models. A very remarkable result of the analysis is the extremely sharp localization in l-space of the spin-dependent part of the S-matrix; as indicated by the smallness of the width parameter Δ_s compared with Δ,

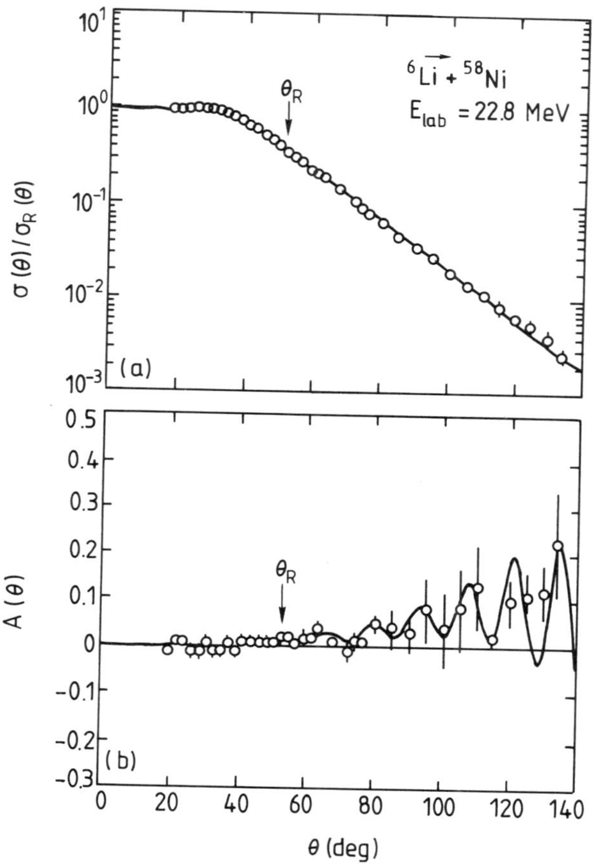

FIG. 6.13. As for Fig. 6.12 but for the system $^{6}\vec{Li}+^{56}Ni$. (From Hill and Frahn 1980 [6.17].)

TABLE 6.6.1
Parameters for $^{6}\vec{Li}$ *scattering and polarization*

Target	n	θ_R (deg)	Λ	Δ	α	Λ_s	Δ_s	κ_R	κ_I
^{28}Si	3.40	29.8	12.9	1.10	0.75	12.06	0.37	0.44	0.14
^{58}Ni	6.81	53.2	13.6	0.88	0	12.56	0.29	0.19	0.04

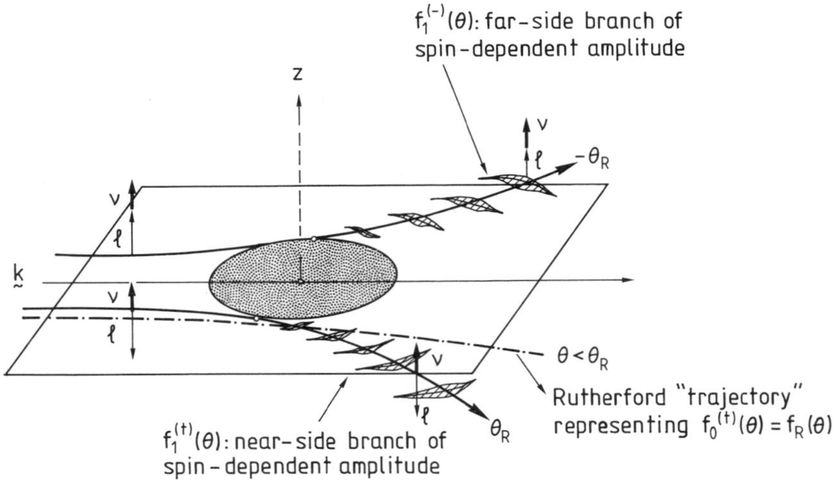

FIG. 6.14. Sketch of the spin-dependent parts of the diffracted waves originating from the near and far-sides of the nucleus. The angular momentum vectors indicate that the relative directions of **l** and **s** are opposite on opposite sides of the nucleus and that, in a coordinate frame with the z axis normal to the scattering plane, the contributions with different z components ν of the spin **s** are scattered independently of one another.

the l-space 'window' associated with the spin–orbit interaction contains effectively only one or two partial waves! The physical reason for this does not yet appear to be understood.

A simple pictorial explanation of the phenomena in spin-dependent heavy-ion scattering is given in Fig. 6.14. We note first that in the semi-quantal approximation and with the choice of quantization axis perpendicular to the scattering plane the total angular momentum of the projectile is $j = l - \nu$ or $j = l + \nu$ depending on whether it is scattered from the near side or the far side of the nucleus. No angular momentum is exchanged between the s components of s and l; the only effect of the spin–orbit coupling is to scatter each component ν of the spin in a different way.

The angular dependence of the vector analysing power is determined by the interference between the spin-independent and spin-dependent amplitudes $f_0(\theta)$ and $f_1(\theta)$:

$$A(\theta) \propto \mathrm{Re}(f_0 f_1^*) = \mathrm{Re}\{(f_0^{(+)} + f_0^{(-)})(f_1^{(+)} + f_1^{(-)})^*\} \tag{6.6.16}$$

Now $f_1(\theta)$, which is given by eqn (6.6.6), has no contribution from a stationary point and can be represented by near-side and far-side diffracted waves as indicated in Fig. 6.14. Under strong Coulomb damping, the scattering in the region $\theta < \theta_R$ is almost entirely near side, and a Fresnel-diffractive pattern arises from the interference between $f_0^{(+)} \approx f_R$ and $f_1^{(+)}$, with $A(\theta)$ oscillating about zero in a similar way as σ/σ_R oscillates about unity (eqn (6.6.14)):

$$A(\theta) \propto \mathrm{Re}(f_R f_1^{(+)*}), \qquad \theta < \theta_R. \tag{6.6.17a}$$

On the other hand, for $\theta > \theta_R$ and strong Coulomb damping in $f_0(\theta)$, the structure of $A(\theta)$ is determined essentially by the interference between the near-side and far-side branches of $f_1(\theta)$:

$$A(\theta) \propto \mathrm{Re}\{f_0^{(+)}(f_1^{(+)} + f_1^{(-)})^*\}. \tag{6.6.17b}$$

This gives rise to a pronounced Fraunhofer-type oscillatory pattern if the θ profiles of $f_1^{(+)}$ and $f_1^{(-)}$, described by the Fourier transforms $\mathscr{P}(\tilde{\theta}_R - \theta)$ and $\mathscr{P}(\tilde{\theta}_R + \theta)$, are relatively *broad* corresponding to a very narrow l-space localization of the spin-dependent part of the S-matrix.

The sensitivity of this structure calls for caution in the interpretation of existing polarization data. The sharp l-space localization inferred from $^6\vec{\mathrm{Li}} + {}^{58}\mathrm{Ni}$ scattering rests on the assumption that the oscillatory structure suggested by the data of $A(\theta)$ for $\theta > \theta_R$ is a real effect; a smoother variation would imply stronger Coulomb damping in $f_1(\theta)$, a narrower θ profile $\mathscr{P}(x)$, and thus a less sharply localized spin part of the S-matrix. In fact, if $\Delta_s \approx \Delta$ both $f_0(\theta)$ *and* $f_1(\theta)$ have strong Coulomb damping, and $A(\theta) \propto \mathrm{Re}(f_0^{(+)} f_1^{(+)*})$ would be smooth in the region $\theta > \theta_R$, as in the example shown in Fig. 6.11. Then there would also be much less enhancement of the scattering cross-section due to the spin contribution $\sigma_1(\theta)$ at large angles.

The effect of the spin–orbit interaction on the large-angle structure of the angular distribution for the elastic scattering of unpolarized heavy ions has also been pointed out recently by Landowne (1981 [6.19]), who thereby explained the observed qualitative differences in the backward-angle cross-sections for $^{13}\mathrm{C} + {}^{28}\mathrm{Si}$ and $^{12}\mathrm{C} + {}^{28}\mathrm{Si}$. The spin-dependent contribution to the cross-section, which is approximately 90° out of phase with the spin-independent contribution, gives a much smoother structure for the pattern for $^{12}\mathrm{C} + {}^{28}\mathrm{Si}$. Moreover, in cases of strong surface absorption a pronounced enhancement of backward scattering by the spin–orbit interaction is predicted, e.g. for $^{13}\mathrm{C} + {}^{40}\mathrm{Ca}$ compared with $^{12}\mathrm{C} + {}^{40}\mathrm{Ca}$.

References

[6.1] Frahn, W. E. (1978). *Nucl. Phys. A* **302,** 281.
[6.2] Rowley, N. (1980). *J. Phys. G* **6,** 697.

[6.3] McIntyre, J. A., Baker, S. D., and Wang, K. H. (1960). *Phys. Rev.* **117,** 1337.
[6.4] Ericson, T. E. D. (1965). In *Preludes in theoretical physics* (eds. A. de Shalit, H. Feshbach, and L. Van Hove), p. 321. North-Holland, Amsterdam.
[6.5] Frahn, W. E. and Venter, R. H. (1963). *Ann. Phys.* (*N.Y*) **24,** 243.
[6.6] Kauffmann, S. K. (1977). *Z. Phys. A* **282,** 163.
[6.7] Hartmann, K. M. (1976). *Nucl. Phys. A* **269,** 237.
[6.8] Frahn, W. E. and Gross, D. H. E. (1976). *Ann. Phys.* (*N.Y*) **101,** 520.
[6.9] Frahn, W. E. (1975). In *Heavy ions, high-spin states and nuclear structure,* Vol. 1, p. 157. International Atomic Energy Agency, Vienna.
[6.10] Blair, J. S. (1954). *Phys, Rev.* **95,** 1218.
[6.11] Frahn, W. E. and Venter, R. H. (1964). *Ann. Phys.* (*N.Y.*) **27,** 135.
[6.12] McIntyre, J. A., Baker, S. D., and Wang, K. H. (1962). *Phys. Rev.* **125,** 584.
[6.13] Weiss, W., Engelhof, P., Hildenbrand, K. D., Kasser, D., Makowska-Rzeszutko, M., Fick, D., Ebinghaus, H., Steffens, E., Amakawa, A., and Kubo, K. I. (1976). *Phys. Lett. B* **61,** 237.
[6.14] Petrovich, F., Stanley, D., Parks, L. A., and Nagel, P. (1978). *Phys. Rev. C* **17,** 1642.
[6.15] Meyer, J., Nahabetian, R. S., and Elbaz, E. (1978). *Nuovo Cimento Lett.* **22,** 355.
[6.16] Fick, D. (1976). *Proc. 4th Int. Symp. on Polarisation Phenomena in Nuclear Reactions* (eds. W. Gruebler and V. Konig), p. 357. Birkhäuser, Basel.
[6.17] Hill, T. F. and Frahn, W. E. (1980). *Ann. Phys.* (*N.Y.*) **124,** 1.
[6.18] Hill, T. F. and Frahn, W. E. (1980). *S. Afr. J. Phys.* **3,** 11.
[6.19] Landowne, S. (1981). *Phys. Rev. C* **23,** 1080.

7

INELASTIC SCATTERING

7.1 DWBA transition amplitude

Inelastic scattering with excitation of low-lying collective states is a process very closely related to elastic scattering. This will be borne out clearly by the formalism developed in this chapter. Our starting point is the partial-wave representation of the inelastic transition amplitude $T_{LM}(\theta)$ for transitions of multipolarity L, M given by the distorted-wave Born approximation (DWBA). The first-order DWBA has proved itself adequate except in the presence of strong coupling between different channels.

Thus we begin with $T_{LM}(\theta)$ in the form

$$T_{LM}(\theta) = \sum_{l_i l_f} \mathrm{i}^{l_i - l_f} \langle l_f L; 00 \mid l_i 0\rangle \langle l_f L; -MM \mid l_i 0\rangle \times (2l_f + 1)^{1/2} R^L_{l_f l_i}(k_f, k_i) \exp[\mathrm{i}\{\sigma_{l_f}(k_f) + \sigma_{l_i}(k_i)\}] Y_{l_f, -M}(\theta, 0) \tag{7.1.1}$$

with the radial integrals

$$R^L_{l_f l_i}(k_f, k_i) = \frac{4\pi}{k_f k_i} \int_0^\infty \mathrm{d}r f^{(f)}_{l_f}(k_f, r) F_L(r) f^{(i)}_{l_i}(k_i, r). \tag{7.1.2}$$

The inelastic form factor is given by the generalized optical model as a sum of two contributions from Coulomb excitation and nuclear excitation:

$$F_L(r) = F^{(C)}_L(r) + F^{(N)}_L(r) = \delta^{(C)}_L Z_1 Z_2 e^2 \frac{3}{2L+1} \frac{R_c^{L-1}}{r^{L+1}} - \delta^{(N)}_L \frac{\mathrm{d}U_N(r)}{\mathrm{d}r}, \tag{7.1.3}$$

where $\delta^{(C)}_L$ and $\delta^{(N)}_L$ are the Coulomb and nuclear deformation lengths, and where the Coulomb term has been taken only for distances larger than the charge radius R_c. The interior part ($r \leq R_c$) of the Coulomb factor $F^{(C)}_L(r)$ is unimportant because for large values of l_i and l_f it has poor overlap with the radial wave functions $f^{(i)}_{l_i}$ and $f^{(f)}_{l_f}$ in the radial integrals (7.1.2) owing to centrifugal repulsion.

Our objective is to evaluate the partial-wave summations in eqn (7.1.1) in analytic form by means of further approximations appropriate under strong-absorption conditions. The resulting expressions will allow us to identify the physical mechanisms operating in inelastic transitions and to display in detail the close connections between inelastic and elastic scattering. In accordance with our general conditions for diffractive processes we assume that the nuclear interactions in the initial and final channels are characterized by S-matrices of normal strong-absorption profile with large values of the grazing angular momenta, $l_0^{(\mathrm{i})} \gg 1$ and $l_0^{(\mathrm{f})} \gg 1$. Then many partial waves contribute to the sums in (7.1.1), the more so because of the long range of the form factor $F_L^{(\mathrm{C})}(r)$ for Coulomb excitation. This allows us, for low angular momentum transfers $L \ll l_0^{(\mathrm{i,f})}$, to replace the Clebsch–Gordan coefficients by their asymptotic forms (Brussard and Tolhoek 1978 [7.1])

$$\langle l_{\mathrm{f}} L; -M M \mid l_{\mathrm{i}} 0\rangle \simeq d^L_{MK}(\tfrac{1}{2}\pi) \tag{7.1.4}$$

in terms of reduced rotation matrix elements, where

$$K = l_{\mathrm{i}} - l_{\mathrm{f}}. \tag{7.1.5}$$

Thus we can write

$$T_{LM}(\theta) = \sum_K \mathrm{i}^K d^L_{0K}(\tfrac{1}{2}\pi) d^L_{MK}(\tfrac{1}{2}\pi) \sum_{l_{\mathrm{f}}} (2l_{\mathrm{f}}+1)^{1/2} b^L_{l_{\mathrm{f}}K}(k_{\mathrm{f}}, k_{\mathrm{i}}) Y_{l_{\mathrm{f}},-M}(\theta, 0), \tag{7.1.6}$$

with the summation over K extending from $-L$ to L, and where

$$b^L_{l_{\mathrm{f}}K}(k_{\mathrm{f}}, k_{\mathrm{i}}) = R^L_{l_{\mathrm{f}}, l_{\mathrm{f}}+K}(k_{\mathrm{f}}, k_{\mathrm{i}}) \exp[\mathrm{i}\{\sigma_{l_{\mathrm{f}}}(k_{\mathrm{f}}) + \sigma_{l_{\mathrm{f}}+K}(k_{\mathrm{i}})\}] \tag{7.1.7}$$

is the partial-wave amplitude for inelastic scattering.

7.2 Inelastic partial-wave amplitude

The main step in our derivation is to use suitable approximations of the radial integrals (7.1.2) for strong absorption in the initial and final channels, which enables us to express the inelastic partial-wave amplitude (7.1.7) in terms of the elastic S-matrices $S_{l_{\mathrm{i}}}(k_{\mathrm{i}})$ and $S_{l_{\mathrm{f}}}(k_{\mathrm{f}})$. For nuclear excitation we employ the Austern–Blair relation with the non-adiabatic modification due to Hahne,

$$R^{(\mathrm{N})L}_{l_{\mathrm{f}} l_{\mathrm{i}}}(k_{\mathrm{f}}, k_{\mathrm{i}}) = \frac{4\pi}{k_{\mathrm{f}} k_{\mathrm{i}}} \int_0^\infty \mathrm{d}r f^{(\mathrm{f})}_{l_{\mathrm{f}}}(k_{\mathrm{f}}, r) F^{(\mathrm{N})}_L(r) f^{(\mathrm{i})}_{l_{\mathrm{i}}}(k_{\mathrm{i}}, r)$$

$$\approx -\mathrm{i}\, \delta^{(\mathrm{N})}_L \frac{\pi \hbar^2}{\mu} \left\{ \frac{\mathrm{d}S^{(\mathrm{N})}_{l_{\mathrm{f}}}(k_{\mathrm{f}})}{\mathrm{d}l_{\mathrm{f}}} \frac{\mathrm{d}S^{(\mathrm{N})}_{l_{\mathrm{i}}}(k_{\mathrm{i}})}{\mathrm{d}l_{\mathrm{i}}} \right\}^{1/2}. \tag{7.2.1}$$

In Appendix 1 it is shown that the square root of the S-matrix derivatives can, to a good approximation, be replaced by the derivative of a K-dependent mean S-matrix. If, as in elastic scattering, the $S_{l_{i,f}}^{(N)}$ are regarded as differentiably continuous functions $S_N(\lambda_{i,f})$ of $\lambda_{i,f}=l_{i,f}+\frac{1}{2}$, we can write

$$R_{l_f l_i}^{(N)L}(k_f,k_i)\approx -i\,\delta_L^{(N)}\frac{\pi\hbar^2}{\mu}\{D_N(\lambda_f)D_N(\lambda_i)\}^{1/2}$$

$$\approx -i\,\delta_L^{(N)}\frac{\pi\hbar^2}{\mu}D_N^{(K)}(\lambda), \qquad (7.2.2)$$

where $D_N(\lambda_{i,f})=dS_N(\lambda_{i,f})/d\lambda_{i,f}$ and $D_N^{(K)}(\lambda)=dS_N^{(K)}(\lambda)/d\lambda$, with

$$S_N^{(K)}(\lambda)=\eta_K(\lambda)\exp\{i2\delta_N^{(K)}(\lambda)\}, \qquad (7.2.3)$$

are defined as functions of the mean angular momentum variable $\lambda=\frac{1}{2}(\lambda_i+\lambda_f)$. The phase functions are defined as

$$2\delta_N^{(K)}(\lambda)=\delta_N^{(i)}(\lambda+\tfrac{1}{2}K)+\delta_N^{(f)}(\lambda-\tfrac{1}{2}K), \qquad (7.2.4)$$

and the reflection functions $\eta_K(\lambda)$ depend on $(\lambda-\Lambda_K)/\Delta$, where

$$\Lambda_K=\Lambda+\Delta_K=\Lambda+\Delta\ln\left\{\cosh\left(\frac{K-K_0}{2\Delta}\right)\right\} \qquad (7.2.5)$$

with

$$\Lambda=\tfrac{1}{2}(\Lambda_i+\Lambda_f), \qquad K_0=\Lambda_i-\Lambda_f. \qquad (7.2.6)$$

In this way we have derived a generalization of the Austern–Blair relation which accounts for the K-dependence and the small nonadiabaticity of the nuclear excitation.

The radial integrals for Coulomb excitation can be approximated by the 'Sopkovich prescription' (Sopkovich 1962 [8.29])

$$R_{l_f l_i}^{(C)L}(k_f,k_i)=\frac{4\pi}{k_f k_i}\int_0^\infty dr f_{l_f}^{(f)}(k_f,r)F_L^{(C)}(r)f_{l_i}^{(i)}(k_i,r)$$

$$\approx \mathscr{R}_{l_f l_i}^{L}(k_f,k_i)\{S_{l_f}^{(N)}(k_f)S_{l_i}^{(N)}(k_i)\}^{1/2}, \qquad (7.2.7)$$

where the $\mathscr{R}_{l_f l_i}^{L}$ are the radial integrals for Coulomb excitation formed with the Coulomb radial wave functions $\mathscr{f}_l(kr)$. These can be written

$$\mathscr{R}_{l_f l_i}^{L}(k_f,k_i)=\delta_L^{(C)}\frac{3}{2L+1}Z_1Z_2e^2R_c^{L-1}\mathscr{M}_{l_f l_i}^{L}(k_f,k_i) \qquad (7.2.8)$$

in terms of the standard integrals of Coulomb excitation theory

$$\mathscr{M}_{l_f l_i}^{L}(k_f,k_i)=\frac{4\pi}{k_f k_i}\int_0^\infty dr\,\mathscr{f}_{l_f}(k_f r)\frac{1}{r^{L+1}}\mathscr{f}_{l_i}(k_i r). \qquad (7.2.9)$$

It is well known (Alder *et al.* 1956 [7.2]) that for $l \gg 1$ these integrals can be evaluated quite accurately in the Wentzel–Kramers–Brillouin (WKB) approximation with the result

$$\mathcal{M}^{L}_{l_f l_i}(k_f, k_i) = \frac{\pi k^{L-2}}{n^L} I_{L-K}(\varepsilon, \xi). \tag{7.2.10}$$

The functions I_{L-K} are defined in Appendix 2 as functions of $\varepsilon = 2\arctan(n/\lambda)$ and the adiabaticity parameter

$$\xi = n_f - n_i, \qquad n_{i,f} = \frac{\mu Z_1 Z_2 e^2}{\hbar^2 k_{i,f}}. \tag{7.2.11}$$

The quantities k, n, and λ in eqn (7.2.10) are the mean values $k = \frac{1}{2}(k_i + k_f)$, $n = \frac{1}{2}(n_i + n_f)$, and $\lambda = \frac{1}{2}(\lambda_i + \lambda_f)$. The approximation (7.2.10) is valid for all values of n if $\lambda \gg 1$.

According to Appendix 1 the square root of the nuclear S-matrices in eqn (7.2.7) can be replaced by the mean S-function (7.2.3). Collecting the results (7.2.2) and (7.2.7)–(7.2.10), we can write the inelastic partial-wave amplitude (7.1.7) as a continuous function of $\lambda = \lambda_f + \frac{1}{2}K$, $b^{L}_{l_f K}(k_f, k_i) \to b_{LK}(\lambda)$, in the form

$$b_{LK}(\lambda) = \frac{\pi\hbar^2}{\mu}(2\lambda)^{1/2} \times \{\delta^{(C)}_{L} c_L I_{L-K}(\varepsilon, \xi) S^{(K)}_{N}(\lambda) - i\delta^{(N)}_{L} D^{(K)}_{N}(\lambda)\}\exp\{i2\sigma^{(K)}(\lambda)\}, \tag{7.2.12}$$

where

$$c_L = \frac{3}{2L+1}\left(\frac{kR_c}{n}\right)^{L-1} \tag{7.2.13}$$

and

$$2\sigma^{(K)}(\lambda) = \sigma^{(i)}(\lambda + \tfrac{1}{2}K) + \sigma^{(f)}(\lambda - \tfrac{1}{2}K). \tag{7.2.14}$$

7.3 'Neutral' particles

As in elastic scattering, it is instructive first to derive expressions for the inelastic scattering amplitude in the absence of Coulomb interaction. Since large-angle inelastic scattering will be treated separately in the next chapter, we can use the 'small-angle' asymptotic form of the spherical harmonics in eqn (7.1.6),

$$Y_{l_f,-M}(\theta, 0) \simeq \left(\frac{\lambda_f}{2\pi}\right)^{1/2}\left(\frac{\theta}{\sin\theta}\right)^{1/2} J_M(\lambda_f \theta), \qquad 0 \leq \theta \lesssim \pi - |M|/\lambda_f, \tag{7.3.1}$$

in terms of cylindrical Bessel functions of order M. By means of the Poisson summation formula, eqn (7.1.6) with $n=0$ can then be written as

$$T_{LM}(\theta) = -\mathrm{i}\delta_L^{(\mathrm{N})}\frac{\pi^{1/2}\hbar^2}{\mu}\left(\frac{\theta}{\sin\theta}\right)^{1/2}\sum_K \beta_{KM}^L \sum_{m=-\infty}^{\infty} \times \exp\{\mathrm{i}2m\pi(\Lambda_K^{\mathrm{f}}-\tfrac{1}{2})\}I_{KM}^{(m)}(\theta), \tag{7.3.2}$$

where

$$\beta_{KM}^L = \mathrm{i}^K d_{0K}^L(\tfrac{1}{2}\pi)d_{MK}^L(\tfrac{1}{2}\pi) \tag{7.3.3}$$

and

$$I_{KM}^{(m)}(\theta) = \int_0^\infty \mathrm{d}\lambda_{\mathrm{f}}\lambda_{\mathrm{f}}D_{\mathrm{N}}^{(K)}(\lambda_{\mathrm{f}}+\tfrac{1}{2}K)J_M(\lambda_{\mathrm{f}}\theta)\exp\{\mathrm{i}2m\pi(\lambda_{\mathrm{f}}-\Lambda_K^{\mathrm{f}})\} \tag{7.3.4}$$

with

$$\Lambda_K^{\mathrm{f}} = \Lambda_K - \tfrac{1}{2}K. \tag{7.3.5}$$

These integrals can be evaluated readily (Frahn 1976 [7.4])

$$I_{KM}^{(m)}(\theta) = \tfrac{1}{2}\Lambda_K^{\mathrm{f}}[F_{\mathrm{N}}\{\Delta(2m\pi-\theta)\}\{J_M(\Lambda_K^{\mathrm{f}}\theta)-\mathrm{i}J_{M+1}(\Lambda_K^{\mathrm{f}}\theta)\} + F_{\mathrm{N}}\{\Delta(2m\pi+\theta)\}\{J_M(\Lambda_K^{\mathrm{f}}\theta)+\mathrm{i}J_{M+1}(\Lambda_K^{\mathrm{f}}\theta)\}], \tag{7.3.6}$$

and we obtain

$$T_{LM}(\theta) = -\mathrm{i}\delta_L^{(\mathrm{N})}\frac{\pi^{1/2}\hbar^2}{2\mu}\left(\frac{\theta}{\sin\theta}\right)^{1/2}\sum_K \beta_{KM}^L\Lambda_K^{\mathrm{f}} \times [G_{KM}(\theta)\{J_M(\Lambda_K^{\mathrm{f}}\theta)-\mathrm{i}J_{M+1}(\Lambda_K^{\mathrm{f}}\theta)\} + G_{KM}(-\theta)\{J_M(\Lambda_K^{\mathrm{f}}\theta)+\mathrm{i}J_{M+1}(\Lambda_K^{\mathrm{f}}\theta)\}], \tag{7.3.7}$$

where

$$G_{KM}(\pm\theta) = \sum_{m=-\infty}^{\infty} \exp\{\mathrm{i}2m\pi(\Lambda_K^{\mathrm{f}}-\tfrac{1}{2})\}F_{\mathrm{N}}\{\Delta(2m\pi\mp\theta)\}. \tag{7.3.8}$$

Expression (7.3.7) is analogous to eqn (6.3.12) for the elastic scattering amplitude and has a similar structure. It consists of two branches, $T_{LM}^{(+)}(\theta)$ and $T_{LM}^{(-)}(\theta)$, corresponding to 'near-side' and 'far-side' inelastic scattering. Again this can be seen from the asymptotic form of the Bessel function combinations:

$$J_M(\Lambda_K^{\mathrm{f}}\theta)\mp\mathrm{i}J_{M+1}(\Lambda_K^{\mathrm{f}}\theta) \simeq \left(\frac{2}{\pi\Lambda_K^{\mathrm{f}}\theta}\right)^{1/2}\exp\{\mp\mathrm{i}(\Lambda_K^{\mathrm{f}}\theta-\tfrac{1}{4}\pi-\tfrac{1}{2}\pi M)\}. \tag{7.3.9}$$

Now, however, there is an additional relative phase $M\pi$ between the (+)

and (−) branches, which expresses the Blair phase rule (Blair 1959 [7.3]) for the Fraunhofer diffraction patterns in inelastic scattering of even and odd multipolarity.

To display the combination of the two amplitude branches we write eqn (7.3.7) in an alternative form by introducing the functions

$$\begin{aligned} \bar{f}_{KM}(\theta) &= \tfrac{1}{2}\{G_{KM}(-\theta)+G_{KM}(\theta)\} \\ \tilde{f}_{KM}(\theta) &= \tfrac{1}{2}\{G_{KM}(-\theta)-G_{KM}(\theta)\}, \end{aligned} \tag{7.3.10}$$

and we obtain

$$\begin{aligned} T_{LM}(\theta) = {} & -\mathrm{i}\delta_L^{(\mathrm{N})}\frac{\pi^{1/2}\hbar^2}{\mu}\left(\frac{\theta}{\sin\theta}\right)^{1/2}\sum_K \beta_{KM}^L \Lambda_K^{\mathrm{f}} \\ & \times\{\bar{f}_{KM}(\theta)J_M(\Lambda_K^{\mathrm{f}}\theta)+\mathrm{i}\tilde{f}_{KM}(\theta)J_{M+1}(\Lambda_K^{\mathrm{f}}\theta)\}. \end{aligned} \tag{7.3.11}$$

The conditions under which only the $m=0$ Poisson terms are significant are essentially the same as those for elastic scattering. To simplify eqn (7.3.11) we neglect the $m\neq 0$ contributions and consider the $K=0$ component only. Further, in order to display the effect of the nuclear phase shifts, let us again adopt the particular form (6.3.17) of the Fourier transform. Then the inelastic differential cross-section can be written as

$$\begin{aligned} \sigma_{LM}(\theta) &= \left(\frac{\mu}{2\pi\hbar^2}\right)^2\frac{k_{\mathrm{f}}}{k_{\mathrm{i}}}|T_{LM}(\theta)|^2 \\ &= \frac{(\delta_L^{(\mathrm{N})})^2}{4\pi}\frac{k_{\mathrm{f}}}{k_{\mathrm{i}}}(\Lambda_0^{\mathrm{f}})^2\{d_{M0}^L(\tfrac{1}{2}\pi)\}^2\frac{\theta}{\sin\theta} \\ &\quad\times(\{J_M(\Lambda_0^{\mathrm{f}}\theta)\}^2+\{\sinh(\alpha\,\Delta\theta)\}^2 \\ &\quad\times[\{J_M(\Lambda_0^{\mathrm{f}}\theta)\}^2+\{J_{M+1}(\Lambda_0^{\mathrm{f}}\theta)\}^2])\{F(\Delta\theta)\}^2. \end{aligned} \tag{7.3.12}$$

In the asymptotic region $\theta \gtrsim |M|/\Lambda_0^{\mathrm{f}}$ this becomes

$$\begin{aligned} \sigma_{LM}(\theta) &\simeq \frac{(\delta_L^{(\mathrm{N})})^2}{2L+1}\{Y_{LM}(\tfrac{1}{2}\pi,0)\}^2\frac{k_{\mathrm{f}}}{k_{\mathrm{i}}}\frac{\Lambda_0^{\mathrm{f}}}{\pi\sin\theta} \\ &\quad\times\{\cosh(2\alpha\,\Delta\theta)+(-)^M\sin(2\Lambda_0^{\mathrm{f}}\theta)\}\{F(\Delta\theta)\}^2 \\ &\approx \frac{(\delta_L^{(\mathrm{N})})^2}{2L+1}\{Y_{LM}(\tfrac{1}{2}\pi,0)\}^2 2\pi\,\Delta^2\Lambda_0^{\mathrm{f}}\frac{\theta^2}{\sin\theta}\exp\{-2(\pi-\alpha)\,\Delta\theta\} \\ &\quad\times\{1+\exp(-4\alpha\,\Delta\theta)+(-)^M 2\exp(-2\alpha\,\Delta\theta)\sin(2\Lambda_0^{\mathrm{f}}\theta)\}. \end{aligned} \tag{7.3.13}$$

Comparison with eqn (6.5.25) shows that the Fourier transforms $F_{\mathrm{N}}(\mp\Delta\theta)$ including the nuclear phase factors have the same effects on the slope and the oscillatory part of the inelastic angular distribution as in the elastic cross-section.

7.4 Charged particles

Now we consider the full partial-wave amplitude (7.2.12) including Coulomb excitation and distortion. Coulomb excitation dominates in inelastic scattering at small angles, and we replace the spherical harmonics by the asymptotic form

$$Y_{l_f,-M}(\theta, 0) \simeq \frac{1}{\pi(\sin\theta)^{1/2}} \cos(\lambda_f\theta - \tfrac{1}{4}\pi - \tfrac{1}{2}\pi M), \qquad \frac{|M|}{\lambda_f} \lesssim \theta \lesssim \pi - \frac{|M|}{\lambda_f}. \tag{7.4.1}$$

Further, under the same conditions as for elastic scattering, only the $m = 0$ term of the Poisson series for the transition amplitude gives a significant contribution at all angles except those close to 180°. Then the inelastic amplitude (7.1.6) has the form

$$\begin{aligned} T_{LM}(\theta) = \sum_K \mathrm{i}^{K+M} d^L_{0K}(\tfrac{1}{2}\pi) d^L_{MK}(\tfrac{1}{2}\pi) &\frac{1}{\pi(2\sin\theta)^{1/2}} \\ \times \int_0^\infty &\mathrm{d}\lambda_f \lambda_f^{1/2} b_{LK}(\lambda_f + \tfrac{1}{2}K) \\ \times &[\exp\{-\mathrm{i}(\lambda_f\theta - \tfrac{1}{4}\pi)\} + (-)^M \exp\{\mathrm{i}(\lambda_f\theta - \tfrac{1}{4}\pi)\}]. \end{aligned} \tag{7.4.2}$$

This amplitude is defined in a coordinate frame with the z-axis in the direction $\mathbf{k}_i$ and the y-axis in the direction $\mathbf{k}_i \times \mathbf{k}_f$. A considerable simplification is obtained by changing to a new coordinate system with the z-axis in the direction $\mathbf{k}_i \times \mathbf{k}_f$ and the x-axis in the direction $\mathbf{k}_i$. This rotation is achieved by the transformation

$$\mathcal{T}_{LM}(\theta) = (-)^M \sum_{M'} \mathrm{i}^{M'} d^L_{M'M}(\tfrac{1}{2}\pi) T_{LM'}(\theta). \tag{7.4.3}$$

Then the summations over M' and K can be carried out by using the symmetry and orthogonality relations of the rotation matrices:

$$\sum_{M'} d^L_{M'M}(\tfrac{1}{2}\pi) d^L_{M'K}(\tfrac{1}{2}\pi) = \delta_{MK} \tag{7.4.4a}$$

$$\sum_{M'} (-)^{M'} d^L_{M'M}(\tfrac{1}{2}\pi) d^L_{M'K}(\tfrac{1}{2}\pi) = (-)^L \delta_{M,-K}. \tag{7.4.4b}$$

The result can be written as

$$\mathcal{T}_{LM}(\theta) = \mathcal{T}^{(+)}_{LM}(\theta) + (-)^M \mathcal{T}^{(-)}_{LM}(\theta), \tag{7.4.5}$$

where

$$\mathcal{T}^{(\pm)}_{LM}(\theta) = \frac{\mathrm{i}^M d^L_{0M}(\tfrac{1}{2}\pi)}{\pi(2\sin\theta)^{1/2}} \int_0^\infty 2\mathrm{d}\lambda_f \lambda_f^{1/2} b_{L,\mp M}(\lambda_f + \tfrac{1}{2}K) \exp\{\mp \mathrm{i}(\lambda_f\theta - \tfrac{1}{4}\pi)\}, \tag{7.4.6}$$

with the understanding that in the rest of this chapter the projection M refers to the normal of the scattering plane as the quantization axis.

For the inelastic scattering amplitudes, redefined as

$$B_{LM}(\theta)=\frac{\mu}{2\pi\hbar^2}\left(\frac{k_\mathrm{f}}{k_\mathrm{i}}\right)^{1/2}T_{LM}(\theta)=B^{(+)}_{LM}(\theta)+(-)^M B^{(-)}_{LM}(\theta), \quad (7.4.7)$$

so that the differential cross-sections are given by

$$\sigma_{LM}(\theta)=|B_{LM}(\theta)|^2, \quad (7.4.8)$$

we obtain, using eqn (7.2.12),

$$\begin{aligned}B^{(\pm)}_{LM}(\theta)=&\frac{y_{LM}}{(2\pi\sin\theta)^{1/2}}\int_0^\infty \mathrm{d}\lambda_\mathrm{f}\lambda_\mathrm{f}^{1/2}\\ &\times\{\delta^{(\mathrm{C})}_L c_L I_{L\pm M}(\varepsilon,\xi)S^{(\mp M)}_\mathrm{N}(\lambda_\mathrm{f}\mp\tfrac{1}{2}M)-\mathrm{i}\delta^{(\mathrm{N})}_L D^{(\mp M)}_\mathrm{N}(\lambda_\mathrm{f}\mp\tfrac{1}{2}M)\}\\ &\times\exp\{\mathrm{i}\phi^{(\mp M)}_\pm(\lambda_\mathrm{f},\theta)\},\end{aligned} \quad (7.4.9)$$

where we have defined

$$y_{LM}=\mathrm{i}^M\left(\frac{k_\mathrm{f}}{k_\mathrm{i}}\right)^{1/2}\frac{d^L_{0M}(\frac{1}{2}\pi)}{(4\pi)^{1/2}}=\mathrm{i}^M\left(\frac{k_\mathrm{f}}{k_\mathrm{i}}\right)^{1/2}\frac{Y_{LM}(\frac{1}{2}\pi,0)}{(2L+1)^{1/2}} \quad (7.4.10)$$

and

$$\phi^{(K)}_\pm(\lambda_\mathrm{f},\theta)=2\sigma^{(K)}(\lambda_\mathrm{f}+\tfrac{1}{2}K)\mp(\lambda_\mathrm{f}\theta-\tfrac{1}{4}\pi). \quad (7.4.11)$$

Separating the contributions from Coulomb and nuclear excitation we write

$$B_{LM}(\theta)=B^{(\mathrm{C})}_{LM}(\theta)+B^{(\mathrm{N})}_{LM}(\theta), \quad (7.4.12)$$

where

$$\begin{aligned}B^{(\mathrm{C},\pm)}_{LM}(\theta)=&\,\delta^{(\mathrm{C})}_L c_L\frac{y_{LM}}{(2\pi\sin\theta)^{1/2}}\\ &\times\int_0^\infty \mathrm{d}\lambda_\mathrm{f}\lambda_\mathrm{f}^{1/2}I_{L\pm M}(\varepsilon,\xi)S^{(\mp M)}_\mathrm{N}(\lambda_\mathrm{f}\mp\tfrac{1}{2}M)\exp\{\mathrm{i}\phi^{(\mp M)}_\pm(\lambda_\mathrm{f},\theta)\}\end{aligned} \quad (7.4.13)$$

$$\begin{aligned}B^{(\mathrm{N},\pm)}_{LM}(\theta)=&-\mathrm{i}\delta^{(\mathrm{N})}_L\frac{y_{LM}}{(2\pi\sin\theta)^{1/2}}\\ &\times\int_0^\infty \mathrm{d}\lambda_\mathrm{f}\lambda_\mathrm{f}^{1/2}D^{(\mp M)}_\mathrm{N}(\lambda_\mathrm{f}\mp\tfrac{1}{2}M)\exp\{\mathrm{i}\phi^{(\mp M)}_\pm(\lambda_\mathrm{f},\theta)\}.\end{aligned} \quad (7.4.14)$$

The integrals in the nuclear excitation components (7.4.14) are easily evaluated. Since the functions $D^{(K)}_\mathrm{N}(\lambda)$ are localized in λ-space over a 'window' of width Δ centred at Λ_K, it suffices to expand the phase functions $\phi_{\pm}^{(K)}(\lambda_{\mathrm{f}}, \theta)$ linearly about $\Lambda_K^{\mathrm{f}} = \Lambda_K - \frac{1}{2}K$:

$$\phi_{\pm}^{(K)}(\lambda_{\mathrm{f}}, \theta) \approx \phi_K(\theta) \pm \tfrac{1}{2}K\theta + (\lambda_{\mathrm{f}} - \Lambda_K^{\mathrm{f}})(\theta_{\mathrm{R}}^{(K)} \mp \theta), \qquad (7.4.15)$$

where

$$\theta_{\mathrm{R}}^{(K)} \equiv \Theta_{\mathrm{R}}^{(K)}(\Lambda_K) = \arctan\left(\frac{n_{\mathrm{i}}}{\Lambda_K + \frac{1}{2}K}\right) + \arctan\left(\frac{n_{\mathrm{f}}}{\Lambda_K - \frac{1}{2}K}\right)$$

$$\approx 2\arctan\left(\frac{n}{\Lambda_K}\right) = \bar{\Theta}_{\mathrm{R}}(\Lambda_K) \qquad (7.4.16)$$

and

$$\phi_K(\theta) \equiv \phi_{\pm}^{(K)}(\Lambda_K, \theta) = 2\sigma^{(K)}(\Lambda_K) \mp (\Lambda_K\theta - \tfrac{1}{4}\pi). \qquad (7.4.17)$$

Thus we obtain

$$B_{LM}^{(\mathrm{N},\pm)}(\theta) = -\mathrm{i}\delta_L^{(\mathrm{N})} y_{LM}(\theta)\kappa_{\mp M}(\theta) F_{\mathrm{N}}\{\Delta(\theta_{\mathrm{R}}^{(\mp M)} \mp \theta)\}\exp\{\mathrm{i}\phi_{\mp M}(\theta)\} \qquad (7.4.18)$$

with

$$\kappa_K(\theta) = \left(\frac{\Lambda_K - \frac{1}{2}K}{2\pi \sin\theta}\right)^{1/2}$$

$$y_{LM}(\theta) = y_{LM}\exp(-\mathrm{i}\tfrac{1}{2}M\theta) = \left(\frac{k_{\mathrm{f}}}{k_{\mathrm{i}}}\right)^{1/2} \frac{Y_{LM}(\frac{1}{2}\pi, 0)}{(2L+1)^{1/2}} \exp\{\mathrm{i}\tfrac{1}{2}M(\pi - \theta)\}, \qquad (7.4.19)$$

and the Fourier transform $F_{\mathrm{N}}(\Delta z)$ defined as in eqn (6.3.11)

$$F_{\mathrm{N}}(\Delta z) = \int_{-\infty}^{\infty} \mathrm{d}\lambda D_{\mathrm{N}}^{(K)}(\lambda)\exp\{\mathrm{i}(\lambda - \Lambda_K)z\} = \Delta\int_{-\infty}^{\infty} \mathrm{d}\mu \hat{D}_{\mathrm{N}}(\mu)\exp(\mathrm{i}\mu\,\Delta z). \qquad (7.4.20)$$

To evaluate the integrals in the Coulomb excitation components (7.4.13) we follow a similar procedure to that described in Section 6.4 for the elastic scattering amplitude. After insertion of an integral representation of $S_{\mathrm{N}}^{(K)}(\lambda)$,

$$S_{\mathrm{N}}^{(K)}(\lambda) = \frac{\mathrm{i}}{2\pi}\lim_{\epsilon\to+0}\int_{-\infty}^{\infty} \mathrm{d}z \frac{\exp\{-\mathrm{i}(\lambda - \Lambda_K)z\}}{z + \mathrm{i}\epsilon} F_{\mathrm{N}}(\Delta z), \qquad (7.4.21)$$

we can write the amplitudes (7.4.13) in the form

$$B_{LM}^{(\mathrm{C},+)}(\theta) = \frac{\mathrm{i}}{2\pi}\lim_{\epsilon\to+0}\int_{-\infty}^{\infty} \mathrm{d}z \frac{F_{\mathrm{N}}(\Delta z)\exp(\mathrm{i}\Lambda_{-M}^{\mathrm{f}}z)}{z + \mathrm{i}\epsilon}\left\{\frac{\sin(z+\theta)}{\sin\theta}\right\}^{1/2} \mathring{B}_{LM}^{(\mathrm{C},+)}(z+\theta) \qquad (7.4.22)$$

$$B_{LM}^{(\mathrm{C},-)}(\theta) = \frac{1}{2\pi}\lim_{\epsilon\to+0}\int_{-\infty}^{\infty} \mathrm{d}z \frac{F_{\mathrm{N}}(\Delta z)\exp(\mathrm{i}\Lambda_M^{\mathrm{f}}z)}{z + \mathrm{i}\epsilon}\left\{\frac{\sin(z-\theta)}{\sin\theta}\right\}^{1/2} \mathring{B}_{L,-M}^{(\mathrm{C},+)}(z-\theta), \qquad (7.4.23)$$

where

$$\mathring{B}^{(\mathrm{C},+)}_{LM}(\vartheta)=\delta^{(\mathrm{C})}_{L}c_L\frac{y_{LM}}{(2\pi\sin\vartheta)^{1/2}}\int_0^\infty \mathrm{d}\lambda_\mathrm{f}\lambda_\mathrm{f}^{1/2}I_{LM}(\varepsilon,\xi)\exp\{\mathrm{i}\phi^{(-M)}_{+}(\lambda_\mathrm{f},\vartheta)\}. \tag{7.4.24}$$

The integral in (7.4.24) can be evaluated in SPA because the non-oscillatory part of the integrand is slowly varying. With the stationary point at $\lambda^\mathrm{f}_\vartheta$ defined by $\Theta^{(-M)}_\mathrm{R}(\lambda^\mathrm{f}_\vartheta-\frac{1}{2}M)=\vartheta$, or at $\lambda_\vartheta=\lambda^\mathrm{f}_\vartheta-\frac{1}{2}M$, this yields

$$\begin{aligned}\mathring{B}^{(\mathrm{C},+)}_{LM}(\vartheta)&=\delta^{(\mathrm{C})}_{L}c_L y_{LM}(\vartheta)\left\{\frac{\lambda_\vartheta+\frac{1}{2}M}{-\bar{\Theta}'_\mathrm{R}(\lambda_\vartheta)\sin\vartheta}\right\}^{1/2}I_{LM}(\vartheta,\xi)\\&\quad\times\exp[\mathrm{i}\{2\sigma^{(-M)}(\lambda_\vartheta)-\lambda_\vartheta\vartheta\}]\\&=\mathrm{i}k\delta^{(\mathrm{C})}_{L}y_{LM}(\vartheta)c_L I_{LM}(\vartheta,\xi)\left(1+\frac{M}{2n}\tan\tfrac{1}{2}\vartheta\right)^{1/2}f_\mathrm{R}(\vartheta)\\&\equiv B^{(\mathrm{CE})}_{LM}(\vartheta),\end{aligned} \tag{7.4.25}$$

where $f_\mathrm{R}(\vartheta)$ is the mean Rutherford scattering amplitude. In fact, eqn (7.4.25) is the WKB approximation of the inelastic scattering amplitude for pure Coulomb excitation, which shall henceforth be denoted by $B^{(\mathrm{CE})}_{LM}(\vartheta)$.

By inserting (7.4.25) in eqns (7.4.22) and (7.4.23) we obtain

$$\begin{aligned}B^{(\mathrm{C},\pm)}_{LM}(\theta)&=\binom{\mathrm{i}}{1}\frac{\delta^{(\mathrm{C})}_{L}c_L y_{LM}(\theta)}{2\pi(\sin\theta)^{1/2}}\exp(\mp\mathrm{i}\Lambda_{\mp M}\theta)\\&\quad\times\lim_{\epsilon\to+0}\int_0^\pi \mathrm{d}\vartheta\left\{\frac{\lambda_\vartheta\pm\frac{1}{2}M}{-\bar{\Theta}'_\mathrm{R}(\lambda_\vartheta)}\right\}^{1/2}\frac{I_{L\pm M}(\vartheta,\xi)F_\mathrm{N}\{\Delta(\vartheta\mp\theta)\}}{\vartheta\mp\theta+\mathrm{i}\epsilon}\\&\quad\times\exp[\mathrm{i}\{2\sigma^{(\mp M)}(\lambda_\vartheta)-(\lambda_\vartheta-\Lambda_{\mp M})\vartheta\}].\end{aligned} \tag{7.4.26}$$

As in the elastic case, the reason for introducing the Fourier transformation (7.4.21) is to obtain an integral representation in which the nonoscillatory part of the integrand is slowly varying, so that the integration over ϑ can be carried out in SPA. The oscillatory parts of the integrands in (7.4.26) have stationary points at the angles

$$\vartheta^{(\pm)}_\mathrm{s}=\bar{\Theta}_\mathrm{R}(\Lambda_{\mp M})=\theta^{(\mp M)}_\mathrm{R}=2\arctan\left(\frac{n}{\Lambda_{\mp M}}\right). \tag{7.4.27}$$

Now we generalize the procedure followed after eqn (6.4.22) in the elastic case by expanding the product $I_{L\pm M}(\vartheta,\xi)F_\mathrm{N}\{\Delta(\vartheta\mp\theta)\}$ about the stationary points $\theta^{(\mp M)}_\mathrm{R}$ *to all orders*. After separating the contributions from the poles near $\vartheta=\pm\theta$, the remaining integrals can be evaluated in SPA, and the summations of the product expansion performed. The result can be written in the form

$$B_{LM}^{(C,\pm)}(\theta)=\mathrm{i}k\delta_L^{(C)}y_{LM}(\theta)c_L I_{L\pm M}(\pm\theta,\xi)f_{\text{sco}}^{(\mp M)}(\theta)$$
$$+\mathrm{i}\delta_L^{(C)}y_{LM}(\theta)\kappa_{\mp M}(\theta)\frac{\exp\{\mathrm{i}\phi_{\mp M}(\theta)\}}{\theta_{\mathrm{R}}^{(\mp M)}\mp\theta}c_L$$
$$\times[I_{L\pm M}(\theta_{\mathrm{R}}^{(\mp M)},\xi)F_{\mathrm{N}}\{\Delta(\theta_{\mathrm{R}}^{(\mp M)}\mp\theta)\}-I_{L\pm M}(\pm\theta,\xi)],\qquad(7.4.28)$$

where

$$f_{\text{sco}}^{(\mp M)}(\theta)=\begin{pmatrix}1\\-\mathrm{i}\end{pmatrix}\frac{\exp(\mp\mathrm{i}\Lambda_{\mp M}\theta)}{2\pi k(\sin\theta)^{1/2}}\lim_{\epsilon\to+0}\int_0^{\pi}\mathrm{d}\vartheta\left\{\frac{\lambda_\vartheta\pm\frac{1}{2}M}{-\bar{\Theta}'_{\mathrm{R}}(\lambda_\vartheta)}\right\}^{1/2}$$
$$\times\frac{\exp[\mathrm{i}\{2\sigma^{(\mp M)}(\lambda_\vartheta)-(\lambda_\vartheta-\Lambda_{\mp M})\vartheta\}]}{\vartheta\mp\theta+\mathrm{i}\epsilon}\qquad(7.4.29)$$

are the branches of the elastic sharp-cut-off amplitudes pertaining to the cut-off angular momenta $\Lambda_{\mp M}$. Using the explicit form of these amplitudes, given by eqns (6.4.32) with $\theta_{\mathrm{R}}\to\theta_{\mathrm{R}}^{(\mp M)}$, we obtain from eqn (7.4.28) together with eqn (7.4.18) our final result for the inelastic scattering amplitude:

$$B_{LM}^{(+)}(\theta)=B_{LM}^{(CE)}(\theta)+\mathrm{i}y_{LM}(\theta)\kappa_{-M}(\theta)$$
$$\times\left[\delta_L^{(C)}c_L I_{LM}(\theta,\xi)\left\{\Gamma_<(\theta_{\mathrm{R}}^{(-M)}-\theta)-\frac{1}{\theta_{\mathrm{R}}^{(-M)}-\theta}\right\}\right.$$
$$\left.+\{\delta_L^{(C)}c_L I_{LM}(\theta_{\mathrm{R}}^{(-M)},\xi)-\delta_L^{(N)}(\theta_{\mathrm{R}}^{(-M)}-\theta)\}\frac{F_{\mathrm{N}}\{\Delta(\theta_{\mathrm{R}}^{(-M)}-\theta)\}}{\theta_{\mathrm{R}}^{(-M)}-\theta}\right]$$
$$\times\exp\{\mathrm{i}\phi_{-M}(\theta)\},\qquad\theta\leqslant\theta_{\mathrm{R}}^{(-M)}\qquad(7.4.30a)$$

$$B_{LM}^{(+)}(\theta)=\mathrm{i}y_{LM}(\theta)\kappa_{-M}(\theta)\left[\delta_L^{(C)}c_L I_{LM}(\theta,\xi)\left\{\Gamma_>(\theta_{\mathrm{R}}^{(-M)}-\theta)-\frac{1}{\theta_{\mathrm{R}}^{(-M)}-\theta}\right\}\right.$$
$$\left.+\{\delta_L^{(C)}c_L I_{LM}(\theta_{\mathrm{R}}^{(-M)},\xi)-\delta_L^{(N)}(\theta_{\mathrm{R}}^{(-M)}-\theta)\}\frac{F_{\mathrm{N}}\{\Delta(\theta_{\mathrm{R}}^{(-M)}-\theta)\}}{\theta_{\mathrm{R}}^{(-M)}-\theta}\right]$$
$$\times\exp\{\mathrm{i}\phi_{-M}(\theta)\},\qquad\theta\geqslant\theta_{\mathrm{R}}^{(-M)}\qquad(7.4.30b)$$

$$B_{LM}^{(-)}(\theta)=\mathrm{i}y_{LM}(\theta)\kappa_M(\theta)\{\delta_L^{(C)}c_L I_{L-M}(\theta_{\mathrm{R}}^{(M)},\xi)-\delta_L^{(N)}(\theta_{\mathrm{R}}^{(M)}+\theta)\}$$
$$\times\frac{F_{\mathrm{N}}\{\Delta(\theta_{\mathrm{R}}^{(M)}+\theta)\}}{\theta_{\mathrm{R}}^{(M)}+\theta}\exp\{\mathrm{i}\phi_M(\theta)\},\qquad(7.4.30c)$$

where $B_{LM}^{(CE)}(\theta)$ is the amplitude for pure Coulomb excitation given by eqn (7.4.25), and where the functions $\Gamma_{\gtrless}(\theta_{\mathrm{R}}^{(-M)}-\theta)$ are defined as in eqns (6.4.33)–(6.4.38). It will be shown in the next section that the two forms (7.4.30a), (7.4.30b) of $B_{LM}^{(+)}(\theta)$ are differentiably continuous at $\theta=\theta_{\mathrm{R}}^{(-M)}$.

(Note that eqns (7.4.30) are generalizations of previously derived expressions (Frahn 1976 [7.4]), in which only the lowest-order terms in the expansion of $I_{L\pm M}(\vartheta, \xi)F_N\{\Delta(\vartheta\mp\theta)\}$ were retained and the M-dependence of the nuclear excitation was neglected.)

7.5 Physical interpretation

7.5.1 *Fresnel diffractive regime*

As in the case of elastic scattering, we first discuss the $\mathscr{C}$-limit or Fresnel-diffractive regime and thereafter the $\mathscr{N}$-limit or Fraunhofer-diffractive regime. Under $\mathscr{C}$-limit conditions, which prevail in the inelastic scattering of heavy ions, it is convenient to consider the ratio of the inelastic amplitude to the amplitude $B_{LM}^{(CE)}(\theta)$ for pure Coulomb excitation. Then eqns (7.4.30) can be rewritten in the form

$$\frac{B_{LM}(\theta)}{B_{LM}^{(CE)}(\theta)} = 1 + c_{-M}(\theta)[(\theta_R^{(-M)} - \theta)\Gamma_<(\theta_R^{(-M)} - \theta) - 1$$
$$+ E_{LM}(\theta)F_N[\Delta(\theta_R^{(-M)} - \theta)\}\{1 + R_{LM}^{(-)}(\theta)\}]$$
$$\times\frac{1}{2\pi^{1/2}u_{-M}}\exp\{-i(u_{-M}^2 + \tfrac{1}{4}\pi)\}\exp\{-i\psi(u_{-M})\}, \qquad \theta\leqslant\theta_R^{(-M)} \tag{7.5.1a}$$

$$\frac{B_{LM}(\theta)}{B_{LM}^{(CE)}(\theta)} = c_{-M}(\theta)[(\theta_R^{(-M)} - \theta)\Gamma_>(\theta_R^{(-M)} - \theta) - 1$$
$$+ E_{LM}(\theta)F_N\{\Delta(\theta_R^{(-M)} - \theta)\}\{1 + R_{LM}^{(-)}(\theta)\}]$$
$$\times\frac{1}{2\pi^{1/2}u_{-M}}\exp\{-i(u_{-M}^2 + \tfrac{1}{4}\pi)\}\exp\{-i\psi(u_{-M})\}, \qquad \theta\geqslant\theta_R^{(-M)} \tag{7.5.1b}$$

with the following definitions:

$$c_{-M}(\theta) = \left(\frac{\sin\theta}{\sin\theta_R^{(-M)}}\right)^{1/2}\frac{\tan\frac{1}{2}\theta}{\tan\frac{1}{2}\theta_R^{(-M)}} \tag{7.5.2}$$

$$E_{LM}(\theta) = \frac{\delta_L^{(C)}c_L I_{LM}(\theta_R^{(-M)}, \xi) - \delta_L^{(N)}(\theta_R^{(-M)} - \theta)}{\delta_L^{(C)}c_L I_{LM}(\theta, \xi)} \tag{7.5.3}$$

$$u_{-M} = \left(\frac{\Lambda_{-M}}{2\sin\theta_R^{(-M)}}\right)^{1/2}(\theta - \theta_R^{(-M)}) = \frac{n^{1/2}}{2\sin\frac{1}{2}\theta_R^{(-M)}}(\theta - \theta_R^{(-M)}). \tag{7.5.4}$$

The function $\psi(u_{-M}) = O(u_{-M}^3)$ is defined in a similar way as in eqn (6.5.3). (Note that in the function $c_{-M}(\theta)$ the small terms $(M/2n)\tan\frac{1}{2}\theta$

and $(M/2n)\tan\frac{1}{2}\theta_{\rm R}^{(-M)}$ have been neglected compared with unity, consistent with retaining only the leading terms in the asymptotic expansion of $f_{\rm sco}^{(\mp M)}(\theta)$.)

The functions $R_{LM}^{(-)}(\theta)$, representing the relative contribution of the far-side branch of the inelastic scattering amplitude, have the form

$$R_{LM}^{(-)}(\theta) = -\mathrm{i}(-)^M\left(\frac{\Lambda_M}{\Lambda_{-M}}\right)^{1/2}\varepsilon_{LM}(\theta)b_M(\theta)\exp\{\mathrm{i}(2\bar{\Lambda}_M\theta + s_M)\}, \quad (7.5.5)$$

where

$$\varepsilon_{LM}(\theta) = \frac{\delta_L^{(\mathrm{C})}c_L I_{L-M}(\theta_{\rm R}^{(M)},\xi) - \delta_L^{(\mathrm{N})}(\theta_{\rm R}^{(M)}+\theta)}{\delta_L^{(\mathrm{C})}c_L I_{LM}(\theta_{\rm R}^{(-M)},\xi) - \delta_L^{(\mathrm{N})}(\theta_{\rm R}^{(-M)}-\theta)} \quad (7.5.6)$$

$$b_M(\theta) = \frac{F_{\rm N}\{\Delta(\theta_{\rm R}^{(M)}+\theta)\}}{\theta_{\rm R}^{(M)}+\theta}\Bigg/\frac{F_{\rm N}\{\Delta(\theta_{\rm R}^{(-M)}-\theta)\}}{\theta_{\rm R}^{(-M)}-\theta} \quad (7.5.7)$$

and

$$\bar{\Lambda}_M = \tfrac{1}{2}(\Lambda_M + \Lambda_{-M}) \quad (7.5.8)$$

$$s_M = 2\sigma^{(M)}(\Lambda_M) - 2\sigma^{(-M)}(\Lambda_{-M}) \approx (\Delta_M - \Delta_{-M})\theta_{\rm R}, \quad (7.5.9)$$

with the Δ_K defined by eqn (7.2.5) and $\theta_{\rm R} \equiv \bar{\Theta}_{\rm R}(\Lambda)$.

Equations (7.5.1*a*) and (7.5.1*b*) show how the structure of the inelastic cross-section $\sigma_{LM}(\theta) = |B_{LM}(\theta)|^2$ differs in the 'illuminated region' $\theta \leqslant \theta_{\rm R}^{(-M)}$ and in the 'shadow region' $\theta \geqslant \theta_{\rm R}^{(-M)}$, which are formally separated by the M-dependent grazing angle

$$\theta_{\rm R}^{(-M)} = 2\arctan\left(\frac{n}{\Lambda_{-M}}\right) \approx \theta_{\rm R} - \frac{\Delta_{-M}}{\Lambda}\sin\theta_{\rm R}. \quad (7.5.10)$$

To study these structures in more detail we first consider the regions not too close to $\theta_{\rm R}^{(-M)}$, such that for $|\theta - \theta_{\rm R}^{(-M)}| \gg (2\Lambda_{-M})^{-1}$ we can use the asymptotic behaviour (6.5.4) of the functions $\Gamma_{\lessgtr}(z)$.

7.5.2 *Illuminated region* $\theta < \theta_{\rm R}^{(-M)}$; *180° phase rule*

For $\theta_{\rm R}^{(-M)} - \theta \gg (2\Lambda_{-M})^{-1}$, eqn (7.5.1*a*) simplifies to

$$\frac{B_{LM}(\theta)}{B_{LM}^{(\mathrm{CE})}(\theta)} \simeq 1 + \frac{c_{-M}(\theta)E_{LM}(\theta)}{2\pi^{1/2}u_{-M}}$$
$$\times\exp\{-\mathrm{i}(u_{-M}^2 + \tfrac{1}{4}\pi)\}\exp\{-\mathrm{i}\psi(u_{-M})\}F_{\rm N}\{\Delta(\theta_{\rm R}^{(-M)}-\theta)\}\{1 + R_{LM}^{(-)}(\theta)\}. \quad (7.5.11)$$

This expression is interpreted in a very similar way as for the corresponding elastic formula (6.5.6). The contributions from the near-side branch give rise to an M-dependent Fresnel diffraction pattern whose oscillations are damped by the function $F_{\mathrm{N}}\{\Delta(\theta_{\mathrm{R}}^{(-M)}-\theta)\}$. This pattern is modulated by oscillations of Fraunhofer-diffractive type arising from the far-side branch and described by the term $R_{LM}^{(-)}(\theta)$. Equation (7.5.5) shows that the angular frequency of these oscillations is $\bar{\Lambda}_M$, for which by means of eqn (7.2.5) we obtain

$$\bar{\Lambda}_M = \Lambda + \tfrac{1}{2}\Delta \ln[\tfrac{1}{2}\{\cosh(M/\Delta) + \cosh(K_0/\Delta)\}], \qquad (7.5.12)$$

and that their amplitudes are damped by the function $b_M(\theta)$ given by eqn (7.5.7).

Apart from the M-dependence, the main difference between the inelastic and elastic Fresnel patterns resides in the function $E_{LM}(\theta)$, given by eqn (7.5.3), which is determined by the relative magnitude of the Coulomb excitation and the nuclear excitation. To show its effect on the angular distribution, we consider conditions of strong Coulomb damping ($|R_{LM}^{(-)}(\theta)| \ll 1$, normally prevailing in heavy-ion collisions). Neglecting $\psi(u_{-M})$ we obtain for the differential cross-section ratio

$$\frac{\sigma_{LM}(\theta)}{\sigma_{LM}^{(\mathrm{CE})}(\theta)} \simeq 1 - E_{LM}(\theta)\frac{c_{-M}(\theta)}{\pi^{1/2}|u_{-M}|}\cos(u_{-M}^2 + \tfrac{1}{4}\pi)F_{\mathrm{N}}[\Delta(\theta_{\mathrm{R}}^{(-M)} - \theta)],$$
$$\theta \lesssim \theta_{\mathrm{R}}^{(-M)} - (2\Lambda_{-M})^{-1}. \qquad (7.5.13)$$

Comparison with eqn (6.5.8) shows that the inelastic Fresnel oscillations would be in phase with the elastic oscillations if $E_{LM}(\theta)$ were always positive. However, eqn (7.5.3) shows that, for angles

$$\theta < \theta_{\mathrm{R}}^{(-M)} - \frac{\delta_L^{(\mathrm{C})}}{\delta_L^{(\mathrm{N})}} c_L I_{LM}(\theta_{\mathrm{R}}^{(-M)}, \xi), \qquad (7.5.14)$$

$E_{LM}(\theta)$ is negative and the inelastic Fresnel oscillations are out of phase with the elastic oscillations. If the nuclear and Coulomb deformation lengths are of comparable magnitude, condition (7.5.14) is satisfied over most of the illuminated region. This *reversal of the inelastic Fresnel pattern owing to the nuclear excitation*, which is independent of the multipolarity L, M, explains the '180° phase rule' observed in heavy-ion angular distributions. It also follows from eqn (7.5.3) that because $I_{LM}(\theta, \xi)$ decreases with decreasing θ, the inelastic Fresnel pattern becomes strongly enhanced over the elastic pattern towards smaller angles.

7.5.3 *Shadow region* $\theta > \theta_R^{(-M)}$

For $\theta - \theta_R^{(-M)} \gg (2\Lambda_{-M})^{-1}$ we obtain from eqn (7.5.1*b*) for the cross-section

$$\sigma_{LM}(\theta) \simeq |y_{LM}|^2 \{\delta_L^{(C)} c_L I_{LM}(\theta_R^{(-M)}, \xi) + \delta_L^{(N)}(\theta - \theta_R^{(-M)})\}^2$$
$$\times \frac{\Lambda_{-M}}{2\pi \sin\theta} \left| \frac{F_N[\Delta(\theta_R^{(-M)} - \theta)]}{\theta_R^{(-M)} - \theta} \right|^2 |1 + R_{LM}^{(-)}(\theta)|^2,$$
$$\theta \gtrsim \theta_R^{(-M)} + (2\Lambda_{-M})^{-1}. \quad (7.5.15)$$

Comparison with eqn (6.5.13) shows that, except for the slow increase of the nuclear part of the excitation factor, the form of the inelastic angular distribution in the shadow region is very similar to the elastic cross-section: the rapid fall-off determined by the Fourier transform $F_N\{\Delta(\theta_R^{(-M)} - \theta)\}$ is smooth except for Fraunhofer-type diffraction oscillations from the far-side contributions $R_{LM}^{(-)}(\theta)$ which are, however, normally strongly damped by the function $b_M(\theta)$ of eqn (7.5.7). The discussion based on the specific example (6.5.14) in the elastic case also applies, with only very minor modifications, to inelastic scattering.

7.5.4 *Near the grazing angle* $\theta_R^{(-M)}$: *the inelastic quarter-point property*

From the definition (6.4.33) of the functions $\Gamma_{\lessgtr}(z)$ it follows that the difference between the right-hand sides of eqns (7.5.1*a*) and (7.5.1*b*) is

$$1 - c_{-M}(\theta)\{1 - (\theta - \theta_R^{(-M)})a_0(u_{-M})\}\exp\{-i\psi(u_{-M})\} = O(\{\theta - \theta_R^{(-M)}\}^2), \quad (7.5.16)$$

showing that our expressions are differentiably continuous at $\theta = \theta_R^{(-M)}$. At this angle we have $E_{LM}(\theta_R^{(-M)}) = 1$, and the amplitude ratio has the value

$$\frac{B_{LM}(\theta_R^{(-M)})}{B_{LM}^{(CE)}(\theta_R^{(-M)})} = \frac{1}{2}\left(1 + \frac{\exp(-i\frac{1}{4}\pi)}{2(\pi n)^{1/2}} \frac{1}{\cos\frac{1}{2}\theta_R^{(-M)}}\right.$$
$$\times [1 + \tfrac{4}{3}(\cos\tfrac{1}{2}\theta_R^{(-M)})^2 - 2\Delta F_N'(0)\sin\theta_R^{(-M)}$$
$$+ i(-)^M \varepsilon_{LM}(\theta_R^{(-M)})\sin\theta_R^{(-M)}$$
$$\left.\times \frac{F_N(2\Delta\theta_R)}{\theta_R} \exp\{i(2\bar{\Lambda}_M\theta_R^{(-M)} + s_M)\}]\right). \quad (7.5.17)$$

This shows that the inelastic scattering cross-section has an approximate quarter-point property relative to the cross-section for pure

Coulomb excitation at the inelastic grazing angle,

$$\frac{\sigma_{LM}(\theta_{\rm R}^{(-M)})}{\sigma_{LM}^{(\rm CE)}(\theta_{\rm R}^{(-M)})}=\frac{1}{4}+O(n^{-1/2}) \tag{7.5.18}$$

up to corrections of order $n^{-1/2}$ which are the same as in the elastic case (eqn (6.5.20)) except for the (normally very small) far-side contribution.

7.5.5 *M-substate population and relative phases*

To display the extent to which our analytic expressions describe the finer features of inelastic scattering we examine the M-dependence of the amplitude $B_{LM}(\theta)$ in more detail by calculating the relative magnitude and relative phase of the ratio

$$\frac{B_{LM}(\theta)}{B_{L-L}(\theta)}=\left|\frac{B_{LM}}{B_{L-L}}\right|\exp\{{\rm i}(\varphi_M-\varphi_{-L})\} \tag{7.5.19}$$

with respect to the reference amplitude for $M=-L$. For simplicity we confine ourselves to the 'asymptotic' Fresnel regime, i.e. we neglect the strongly damped far-side branches and consider only the regions $\theta\lesssim\theta_{\rm R}^{(-M)}-(2\Lambda_{-M})^{-1}$ and $\theta\gtrsim\theta_{\rm R}^{(-M)}+(2\Lambda_{-M})^{-1}$. As nothing special happens at or near $\theta_{\rm R}^{(-M)}$ this is not a significant restriction.

Apart from the quantities $y_{LM}(\theta)$ defined in (7.4.19), the main M-dependence comes from the Coulomb excitation functions $I_{LM}(\theta,\xi)$. Compared with this the M-dependence of the nuclear excitation is weak, the relative differences in Λ_{-M} and $\theta_{\rm R}^{(-M)}$ being of order Δ/Λ. To simplify the discussion still further we neglect these differences and replace Λ_{-M} and $\theta_{\rm R}^{(-M)}$ by the mean values Λ and $\theta_{\rm R}$.

In the illuminated region we derive from eqn (7.5.11)

$$\frac{B_{LM}(\theta)}{B_{L-L}(\theta)}\approx\frac{y_{LM}(\theta)I_{LM}(\theta,\xi)}{y_{L-L}(\theta)I_{L-L}(\theta,\xi)}\times\left[1-E_{LM}(\theta)\frac{c(\theta)}{2\pi^{1/2}|u|}\exp\{-{\rm i}(u^2+\tfrac{1}{4}\pi)\}F_{\rm N}\{\Delta(\theta_{\rm R}-\theta)\}\right], \tag{7.5.20}$$

where we have used the fact that in the $M=-L$ substate the amplitude of the damped Fresnel oscillations is small compared with unity (at most 0.1 in the asymptotic region), and where we have neglected $E_{L-L}(\theta)$ compared with $E_{LM}(\theta)$ because $I_{LM}(\theta,\xi)$ with $M\neq-L$ is much larger than $I_{L-L}(\theta,\xi)$ at small angles. With

$$\frac{y_{LM}(\theta)}{y_{L-M}(\theta)}=\frac{L!}{(2L!)^{1/2}}\frac{\{(L+M)!(L-M)!\}^{1/2}}{\{(L+M)/2\}!\{(L-M)/2\}!}\times\exp\{-{\rm i}\tfrac{1}{2}(L+M)\theta\}\equiv\zeta_{LM}\exp\{-{\rm i}\tfrac{1}{2}(L+M)\theta\}, \tag{7.5.21}$$

we obtain

$$\left|\frac{B_{LM}}{B_{L-L}}\right| \approx \zeta_{LM} \frac{I_{LM}(\theta, \xi)}{I_{L-L}(\theta, \xi)} [1+\{a_{LM}(\theta)\}^2 + 2a_{LM}(\theta)\cos(u^2+\tfrac{1}{4}\pi)]^{1/2} \tag{7.5.22}$$

and

$$\varphi_M - \varphi_{-L} \approx -\tfrac{1}{2}(L+M)\theta - \arctan\left\{\frac{a_{LM}(\theta)\sin(u^2+\frac{1}{4}\pi)}{1+a_{LM}(\theta)\cos(u^2+\frac{1}{4}\pi)}\right\}, \tag{7.5.23}$$

where

$$a_{LM}(\theta) = -E_{LM}(\theta) \frac{c(\theta)}{2\pi^{1/2}|u|} F_N\{\Delta(\theta_R - \theta)\}. \tag{7.5.24}$$

Expressions (7.5.22) and (7.5.23) display a number of interesting features.

(i) The relative magnitude of the amplitude B_{LM} is mainly determined by the Coulomb excitation functions $I_{LM}(\theta, \xi)$, whose well-known properties are summarized in Appendix 2. Since $I_{LM}(\theta, \xi)$ is largest for $M=-L$, smaller for $M=-L+2$ and so on, and smallest for $M=L$, the magnitude of the substate cross-section $\sigma_{LM}(\theta)$ depends strongly on M, especially at small angles where

$$\sigma_{L-L}(\theta) \gg \sigma_{L,-L+2}(\theta) \gg \cdots \gg \sigma_{LL}(\theta). \tag{7.5.25}$$

(ii) The oscillatory part of each $|B_{LM}(\theta)|$ and $\sigma_{LM}(\theta)$ is a Fresnel pattern, damped by the function $F_N\{\Delta(\theta_R-\theta)\}$. These patterns are all in phase, in accordance with the 180° phase rule. However, since $E_{LM}(\theta)$ is inversely proportional to $I_{LM}(\theta, \xi)$, the amplitude $a_{LM}(\theta)$ of the Fresnel oscillations is very different for different M: it is smallest for $M=-L$ and largest for $M=L$. In fact, all Coulomb–nuclear interference effects become more pronounced with increasing M, especially at smaller angles.

(iii) The relative phases (7.5.23) consist of a term linear in M, which represents the 'adiabatic limit' for negligible Coulomb excitation (Blair and Wilets 1961 [7.5]; Cramer and Eidson 1964 [7.6]), and an oscillatory part which is again of the Fresnel-diffractive type. The latter has the same properties as the pattern in $|B_{LM}(\theta)|$; in particular its amplitude increases strongly with increasing M, except for a 90° phase difference. This can be formulated as a '90° phase rule' (Frahn 1976 [7.4]): the Fresnel pattern of the relative phases is (independent of the multipolarity) shifted by +90° relative to that of the inelastic cross-section, and by −90° relative to that of the elastic cross-section.

(iv) Additional structure can occur in $\sigma_{LM}(\theta)$ and in $\varphi_M - \varphi_{-L}$ if the magnitude of $a_{LM}(\theta)$ becomes comparable with unity. (Recall that

$E_{LM}(\theta)$ is negative, and thus $a_{LM}(\theta)$ is positive, over most of the illuminated region.) This can happen for high M, especially for $M=L$, if the Fresnel amplitude $a_{LM}(\theta)$ is sufficiently enhanced by the nuclear phase factor contained in the Fourier transform $F_{\mathrm{N}}\{\Delta(\theta_{\mathrm{R}}-\theta)\}$. Then $\sigma_{LM}(\theta)$ shows a *sharp dip* and $\varphi_M-\varphi_{-L}$ shows a *sharp dispersion-like structure* at the angle θ_0 where $u^2=\frac{3}{4}\pi$, so that $\cos(u^2+\frac{1}{4}\pi)=-1$. From the discussion in Section 6.5.1 we recall that θ_0, given by eqn (6.5.10), is the location of the main peak of the elastic Fresnel pattern, and eqn (7.5.13) shows that θ_0 is also the position of the main minimum in the inelastic Fresnel pattern which becomes progressively deeper as M increases from $-L$ to $+L$.

Turning to the shadow region $\theta \gtrsim \theta_{\mathrm{R}}+(2\Lambda)^{-1}$, we obtain from eqn (7.5.1b)

$$\left|\frac{B_{LM}(\theta)}{B_{L-L}(\theta)}\right| \simeq \zeta_{LM}\frac{\delta_L^{(\mathrm{C})}c_L I_{LM}(\theta_{\mathrm{R}},\xi)+\delta_L^{(\mathrm{N})}(\theta-\theta_{\mathrm{R}})}{\delta_L^{(\mathrm{C})}c_L I_{L-L}(\theta_{\mathrm{R}},\xi)+\delta_L^{(\mathrm{N})}(\theta-\theta_{\mathrm{R}})} \tag{7.5.26}$$

and

$$\varphi_M-\varphi_{-L}\simeq -\tfrac{1}{2}(L+M)\theta. \tag{7.5.27}$$

Thus the amplitude ratios and substate cross-sections are nonoscillatory and their magnitudes depend only weakly on M. The relative phases for $\theta \geqslant \theta_{\mathrm{R}}$ vary smoothly with θ and soon approach the asymptotic 'adiabatic limit' (7.5.27).

All the features described in this subsection are strikingly illustrated by Figs. 7.1 and 7.2, which show the data of Steadman *et al.* (1974 [7.7]) for the inelastic scattering of ^{16}O by ^{56}Fe at $E_{\mathrm{lab}}=43$ MeV with excitation of the 847 keV 2^+ state in ^{56}Fe. The curves are calculated using the closed formalism (Frahn and Rehm 1976, 1978 [7.8, 7.9]).

7.5.6 *Fraunhofer diffractive regime*

Under conditions where the Coulomb distortion is weak and the Coulomb excitation is negligibly small compared with the nuclear excitation (i.e. for light particles or higher energies or both), we obtain from eqns (7.4.30b) and (7.4.30c) with $\delta_L^{(\mathrm{C})}=0$,

$$\begin{aligned} B_{LM}(\theta) = -\mathrm{i}\delta_L^{(\mathrm{N})} y_{LM}(\theta)[&\kappa_{-M}(\theta)F_{\mathrm{N}}\{\Delta(\theta_{\mathrm{R}}^{(-M)}-\theta)\}\exp\{\mathrm{i}\phi_{-M}(\theta)\} \\ &+(-)^M\kappa_M(\theta)F_{\mathrm{N}}\{\Delta(\theta_{\mathrm{R}}^{(M)}+\theta)\}\exp\{\mathrm{i}\phi_M(\theta)\}], \qquad \theta\geqslant\theta_{\mathrm{R}}^{(-M)}. \end{aligned} \tag{7.5.28}$$

Comparison with eqn (6.5.26) for the elastic amplitude shows that, apart from the weak M-dependence of the grazing angle, the effects of the

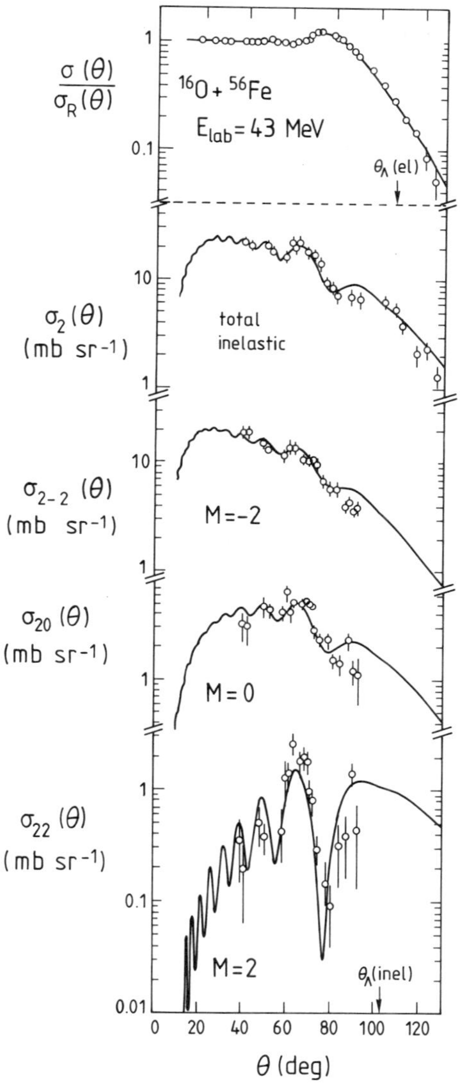

FIG. 7.1. Angular distributions for the elastic and inelastic scattering of $^{16}O+^{56}Fe$ at $E_{lab}=43$ MeV. The curves are closed-formalism fits to the data of Steadman *et al.* (1978 [7.7]), with parameters based on a best fit to the total inelastic cross-section.

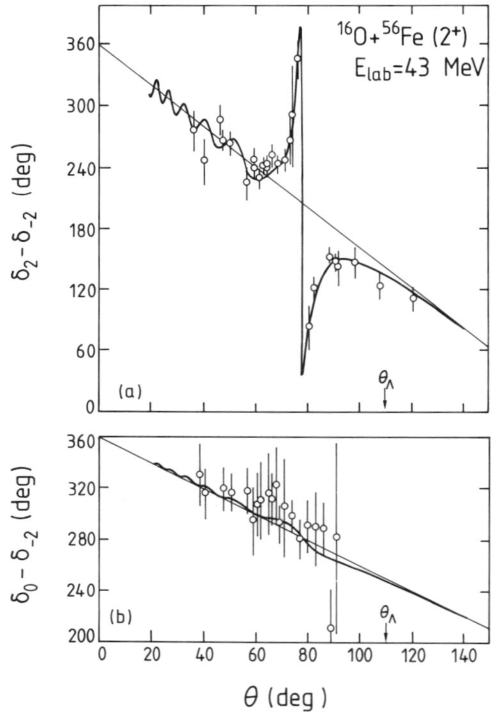

FIG. 7.2. Relative phases (a) $\delta_2-\delta_{-2}$ and (b) $\delta_0-\delta_{-2}$ of the inelastic scattering amplitudes B_{2M} for the reaction $^{18}O+{}^{56}Fe$ (2+) at $E_{lab}=43$ MeV. The curves are closed-formalism fits to the data of Steadman *et al.* (1978 [7.7]) with parameters based on a best fit to the total inelastic cross-section.

damping functions $F_N\{\Delta(\theta_R\mp\theta)\}$ are essentially the same in inelastic and elastic scattering.

In the extreme $\mathcal{N}$-limit, and neglecting the small difference between Λ_{-M} and Λ_M, eqn (7.5.28) reduces to

$$B_{LM}(\theta)=-\mathrm{i}\delta_L^{(N)}y_{LM}(\theta)\left(\frac{\Lambda}{2\pi\sin\theta}\right)^{1/2}$$
$$\times[F_N(-\Delta\theta)\exp\{-\mathrm{i}(\Lambda\theta-\tfrac{1}{4}\pi)\}+(-)^M F_N(\Delta\theta)\exp\{\mathrm{i}(\Lambda\theta-\tfrac{1}{4}\pi)\}], \tag{7.5.29}$$

which should be compared with eqn (6.5.27) for the elastic amplitude. It follows that the relative magnitude of the M-substate amplitudes and their relative phases in the $\mathcal{N}$-limit are given by

$$\left|\frac{B_{LM}}{B_{L-L}}\right| = \zeta_{LM}, \qquad \varphi_M - \varphi_{-L} = -\tfrac{1}{2}(L+M)\theta, \tag{7.5.30}$$

in agreement with the corresponding limit of eqns (7.5.26) and (7.5.27).

The differential cross-section calculated from eqn (7.5.29),

$$\sigma_{LM}(\theta) = (\delta_L^{(\mathrm{N})} y_{LM})^2 \frac{\Lambda}{2\pi \sin\theta} \{F_{\mathrm{N}}(-\Delta\theta)\}^2$$
$$\times [1 + \{b(\theta)\}^2 + 2(-)^L b(\theta) \sin(2\Lambda\theta)] \tag{7.5.31}$$

where $b(\theta) = F_{\mathrm{N}}(\Delta\theta)/F_{\mathrm{N}}(-\Delta\theta)$, shows explicitly the Blair phase rule for the inelastic Fraunhofer pattern.

7.5.7 *Pictorial explanation*

Because of the close analogies between inelastic and elastic scattering displayed by our analytic expressions, the main features of the inelastic angular distributions are explained in much the same way as in the elastic case. The corresponding process to Rutherford scattering is pure Coulomb excitation: the amplitude $B_{LM}^{(\mathrm{CE})}(\theta)$ is the analogue of $f_{\mathrm{R}}(\theta)$. Since Coulomb excitation is a long-range interaction it can be regarded as taking place along a classical trajectory; this finds its mathematical expression in evaluating the corresponding part of the inelastic amplitude in SPA. In contrast, for nuclear excitation under strong absorption conditions the trajectory picture is inappropriate. Because of the short range of the nuclear interaction, the nuclear excitation proceeds through a narrow window of width Δ in angular momentum space. The l-space-localized profile of the nuclear part of the inelastic partial-wave amplitude (7.2.12) is described by the absorptive shape function $D_{\mathrm{N}}(\lambda)$, and the predominantly diffractive nature of the nuclear excitation manifests itself in the appearances of the Fourier transforms $F_{\mathrm{N}}(\Delta z)$ in our formulae. Since these are the same functions as in the elastic amplitude, they have the same effect on the inelastic cross-section–Coulomb damping in the Fresnel regime and refractive damping in the Fraunhofer regime—as was described in Section 6.5.5 for the elastic case.

The occurrence of Fresnel-diffractive patterns in the inelastic angular distributions can also be explained as due to a 'diverging lens' effect of the Coulomb field in the presence of strong absorption. The phase reversal of the inelastic Fresnel oscillations relative to the elastic oscillations as expressed by the 180° phase rule, comes about because of the destructive interference between nuclear and Coulomb excitation.

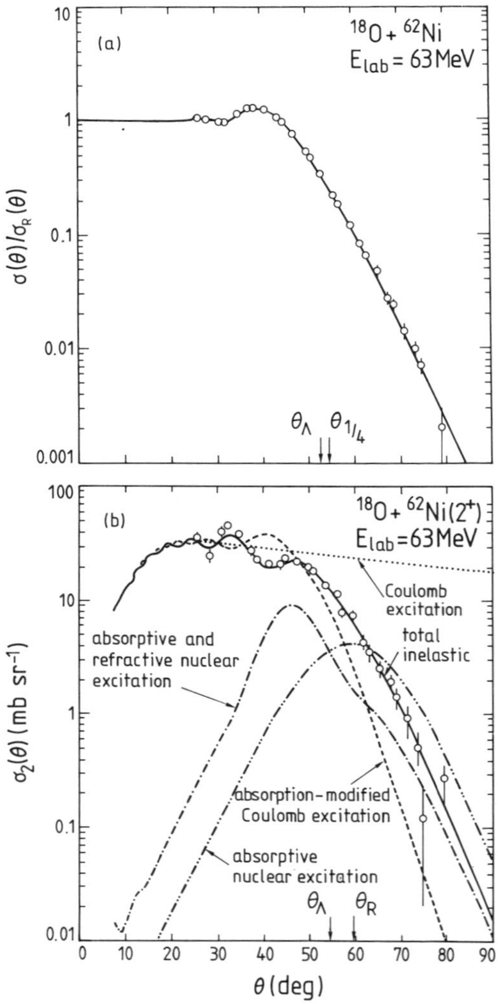

FIG. 7.3. Differential cross-sections for the elastic and inelastic scattering of ^{18}O by ^{62}Ni showing the contributions of pure Coulomb excitation, absorption modified Coulomb excitation and nuclear excitation to the inelastic scattering in the Fresnel region.

The main differences between inelastic and elastic scattering are due to the effects of the multipolarity L, M. The weak M-dependence of the effective grazing angular momenta $\Lambda_{\mp M}$ and associated grazing angles $\theta_R^{(\mp M)}$ arises from the small kinematical shifts in the mean partial-wave S-matrix (7.2.3). As we have seen, the main dependence on multipolarity comes from the Coulomb excitation functions $I_{LM}(\theta, \xi)$ and from the relative phases of the oscillatory parts of the amplitude. The latter give rise to two different types of phase rule: the 180° and 90° phase rules in the Fresnel-diffractive regime, and the Blair phase rule for the Fraunhofer oscillations. That the 180° phase reversal is independent of the multipolarity is explained by the Fresnel oscillations arising from interference within the near-side branch of the amplitude only, whereas the Fraunhofer patterns alternate in sign for even and odd multipolarity because of interference between the near-side and far-side branches.

In Fig. 7.3 we show an example of how the inelastic cross-section in the Fresnel regime is composed of its various contributions: pure Coulomb excitation, absorption-modified Coulomb excitation, and nuclear excitation. It is seen how the Fresnel oscillations in the curve for absorption-modified Coulomb excitation, which are *in* phase with the elastic Fresnel pattern, are reversed by interference with the nuclear excitation. A striking illustration of the 180° phase rule is shown in Fig. 7.4, which

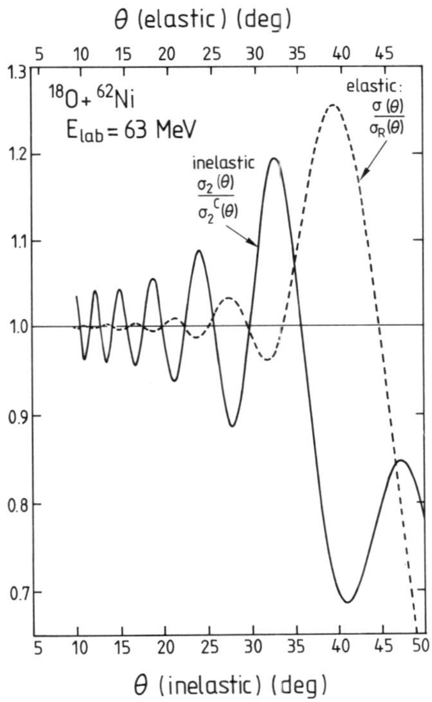

FIG. 7.4. Comparison of the cross-section ratios $\sigma_2(\theta)/\sigma_2^{(CE)}(\theta)$ and $\sigma(\theta)/\sigma_R(\theta)$ showing the 180° phase rule and the enhancement of the inelastic Fresnel oscillations relative to the elastic ones towards smaller angles.

gives comparison of the cross-section ratios $\sigma_2(\theta)/\sigma_2^{(CE)}(\theta)$ and $\sigma(\theta)/\sigma_R(\theta)$. This also depicts the enhancement of the inelastic Fresnel oscillations relative to the elastic oscillations towards smaller angles. As a further example, a comparison between elastic scattering and inelastic scattering of different (even and odd) multipolarities is shown in Fig. 7.5. Finally, Fig. 7.6 gives an example of elastic and inelastic scattering in the Fraunhofer-diffractive regime.

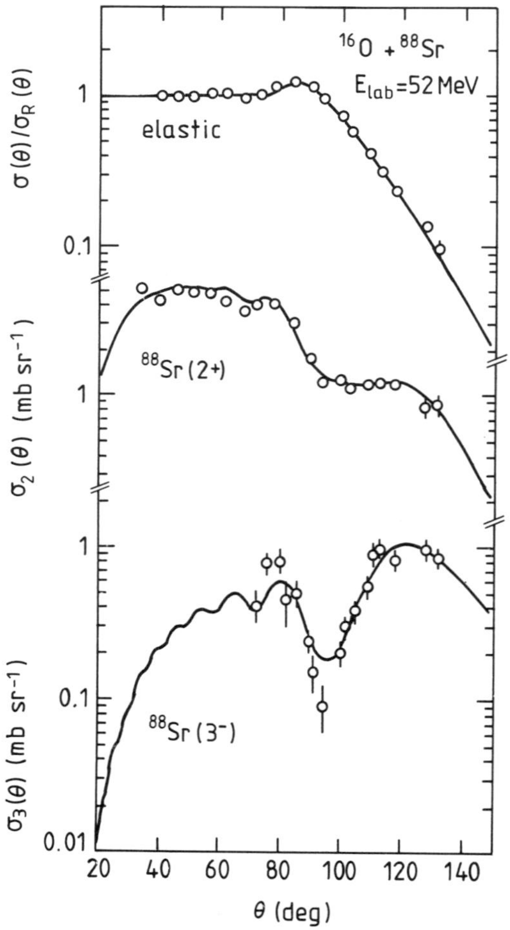

FIG. 7.5. Differential cross-sections for the elastic and inelastic scattering of ^{16}O by ^{88}Sr showing different multipolarities.

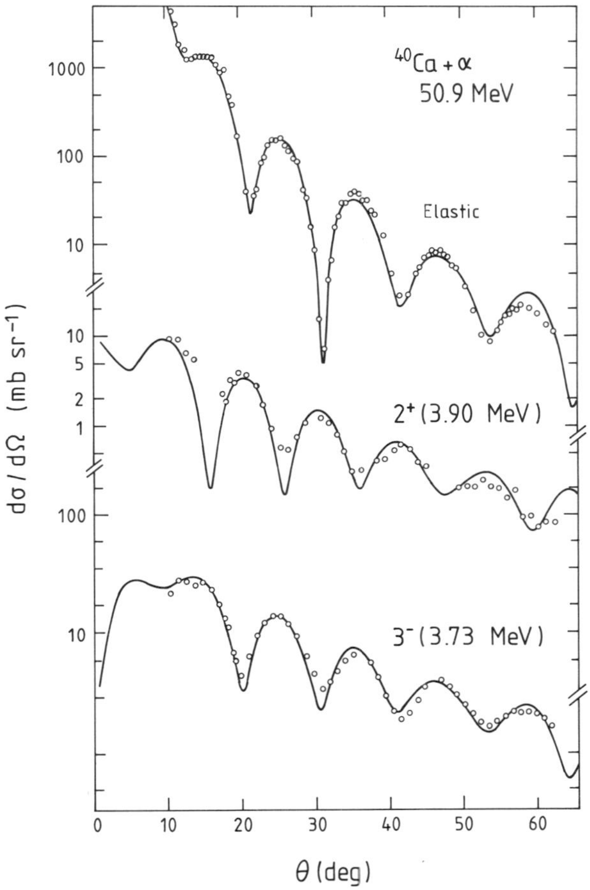

FIG. 7.6. Differential cross-sections for the elastic and inelastic scattering of alpha-particles by ^{40}Ca in the Fraunhofer diffractive regime.

7.6 References

[7.1] Brussard, P. J. and Tolhoek, H. A. (1958). *Physica* **24,** 233.

[7.2] Alder, K., Bohr, A., Huus, T., Mottelson, B., and Winther, A. (1956). *Rev. mod. Phys.* **28,** 432.

[7.3] Blair, J. S. (1959). *Phys. Rev.* **115,** 928.

[7.4] Frahn, W. E. (1976). *Nucl. Phys. A* **272,** 413.

[7.5] Blair, J. S. and Wilets, L. (1961). *Phys. Rev.* **121,** 1493.

[7.6] Cramer, J. G. and Eidson, W. W. (1964). *Nucl. Phys.* **55,** 593.

[7.7] Steadman, S. G., Belote, T. A., Goldstein, R., Grodzins, L., Cline, D., de Vorgt, M. J. A., and Videbaek, F. (1974). *Phys. Rev. Lett.* **33,** 449, 1320.

[7.8] Frahn, W. E. and Rehm, K. E. (1976). *Lett. nuovo Cimento* **17,** 339.

[7.9] Frahn, W. E. and Rehm, K. E. (1978). *Phys. Rep.* **37C,** 1.

8

TRANSFER REACTIONS

8.1 Introduction

In this chapter we give an analytic description of 'quasi-elastic' (as distinct from 'deeply inelastic') transfer reactions between heavy ions, based on distorted-wave theory. The purpose is to display explicitly all significant features of the transition amplitude, including the finer details of the dependence on magnetic quantum numbers, l-matching, Q-values, etc.

The earliest closed-form treatments of heavy-ion transfer reactions above the Coulomb barrier (Frahn and Venter 1964 [8.1]; Strutinsky 1964, 1973 [8.2, 8.3]) assumed simple phenomenological forms of the partial-wave amplitude and were restricted to $L=0$ transfer. Although grossly oversimplified, these descriptions emphasized the predominantly diffractive nature of the transfer process and were the first to predict the existence of regular oscillations in the transfer cross-section at small angles, even though the data known at the time showed smooth 'bell-shaped' angular distributions only. Also, the systematic variations of the main cross-section characteristics with energy, charge, and mass numbers were at least qualitatively described. At first, however, the diffraction oscillations were not found at the predicted energies, which was attributed (Dodd and Greider 1965, 1969 [8.4, 8.5]) to the neglect of recoil effects. The oscillatory structure of many transfer angular distributions was not firmly established by experiment until early in the 1970s (Christensen *et al.* 1973 [8.6]; Bond *et al.* 1973 [8.7]; Lemaire *et al.* 1979 [8.8]; LeVine *et al.* 1974 [8.9]; Henning *et al.* 1974 [8.10]).

Closed-form descriptions based on the DWBA were first developed by Dar and coworkers (Dar 1965, 1966 [8.11, 8.12]; Dar and Kozlowsky 1965 [8.13]; Varma 1968 [8.14]) who used the parametric form of the strong absorption model (Frahn and Venter 1963 [8.15]) for the mean elastic S-matrix. This formulation was extended (Frahn and Sharaf 1969 [8.16]) by employing a more general technique for the partial-wave summation.

A different approach, with emphasis on the classical refractive rather than the quantal diffractive aspects of heavy-ion reactions, was followed in the early semiclassical theory (Broglia and Winther 1972 [8.17]). This was later amended to account more appropriately for the effects of strong absorption by the use of complex trajectories and saddle-point methods (Landowne *et al.* 1976 [8.18]; Knoll and Schaeffer 1977 [8.19]). The most recent semiclassical treatment (Hasan and Brink 1978 [8.20]), although similar in concept to that of Landowne *et al.* (1976 [8.18]), takes a significant step towards a closed-form representation by performing analytically the main partial-wave summation in the distorted-wave transition amplitude.

Much of the recent advance in a more detailed understanding of heavy-ion transfer reactions is due to the experimental and theoretical work of Kahana and coworkers (Chasman *et al.* 1973 [8.21]; Schneider *et al.* 1973 [8.22]; Kahana *et al.* 1974 [8.23]). These workers have interpreted several significant features of the distorted-wave amplitude that were not accounted for in the earlier closed-form descriptions by means of an extended version of the phenomenological transfer S-matrix (Strutinsky 1964, 1973 [8.2, 8.3]) in which the localization with respect to the difference between the initial and final orbital angular momenta is incorporated explicitly. A clear and useful presentation of these developments is given in the review by Kahana and Baltz (1977 [8.24]).

In this chapter we employ analytic techniques developed previously for elastic (Frahn 1975 [8.25, 8.26]) and inelastic (Frahn 1976 [8.27]) heavy-ion scattering to evaluate the partial-wave representation of the transfer transition amplitude. Particular attention is given to a more accurate treatment of the small- and large-angle regions, and to showing how the specific properties pointed out by Kahana and coworkers can be derived from the transfer S-matrix of distorted-wave theory.

8.2 Small-angle representation

8.2.1 *Angular distribution*

We assume that the reaction

$$\mathrm{A_i+B_i=(A_f+c)+B_i\rightarrow A_f+(B_i+c)=A_f+B_f}, \qquad (8.2.1)$$

in which a nucleon or cluster c is transferred between the (heavy) nuclei $\mathrm{A_i}$ and $\mathrm{B_i}$, can be described using the DWBA. In view of excellent accounts of basic distorted-wave theory given by Kahana and Baltz (1977 [8.24]) and Austern (1970 [8.28]) we can start directly from the partial-wave representation of the DWBA transfer amplitude. With all due reservations, particularly for heavy ions, we adopt the no-recoil approximation for simplicity, leaving recoil corrections for consideration at a later stage. Then the differential cross-section for angular momentum transfer L can be written

$$\sigma_L^{(T)}(\theta) = A\frac{\mu_i\mu_f}{(2\pi\hbar^2)^2}\frac{k_f}{k_i}\sum_{M=-L}^{L}|\beta_{LM}(\theta)|^2, \tag{8.2.2}$$

where μ_i and μ_f are the reduced masses and k_i and k_f the relative wave numbers in the initial and final channels. We have disregarded spin effects and lumped all spin and spectroscopic factors together in the constant A. Then the reduced transition amplitude for multipolarity (L, M) is given by

$$\beta_{LM}(\theta) = \sum_{l_il_f} i^{l_i-l_f-L}\langle l_fL; 00 \mid l_i0\rangle\langle l_fL; -MM \mid l_i0\rangle$$
$$\times (2l_f+1)^{1/2}R^L_{l_fl_i}(k_f, k_i)\exp[i\{\sigma_{l_f}(k_f)+\sigma_{l_i}(k_i)\}]Y_{l_f,-M}(\theta, 0). \tag{8.2.3}$$

Here we have chosen a coordinate frame with the z-axis in the direction of $\mathbf{k}_i$ and the y-axis in the direction of $\mathbf{k}_i \times \mathbf{k}_f$. The radial integrals are defined as

$$R^L_{l_fl_i}(k_f, k_i) = \frac{4\pi}{\zeta k_f k_i}\int_0^\infty dr f^{(f)}_{l_f}(k_f, \zeta r)F^{(\kappa)}_L(r)f^{(i)}_{l_i}(k_i, r), \tag{8.2.4}$$

where $f_l^{(i,f)}$ are the radial parts of the distorted wave functions, and

$$F^{(\kappa)}_L(r) = -i^L h^{(1)}_L(i\kappa r) \simeq \frac{\exp(-\kappa r)}{\kappa r} \tag{8.2.5}$$

is the transfer form factor; $\kappa = (2\mu_T\varepsilon_B)^{1/2}/\hbar$ is the imaginary wave number associated with the binding energy ε_B of the transferred cluster in the final nucleus. The scale factor ζ arises from the no-recoil approximation and can vary between $\zeta = B_i/B_f$ and $\zeta = (B_i/B_f)\{1+(c/\mu_i)\}$ depending on the range of the interaction (Kahana and Baltz 1977 [8.24]) (where the symbols of the particles stand for their masses), and the $\sigma_l(k)$ denote the Rutherford phase shifts.

Our purpose is to evaluate the summations over the orbital angular momenta l_i, l_f in explicit analytic form by utilizing the characteristic simplifying features of heavy-ion interactions above the Coulomb barrier. These are (i) large grazing angular momenta $l_i^{(0)}$, $l_f^{(0)}$, (ii) strong Coulomb interaction indicated by high values of the Sommerfeld parameters n_i, n_f, and (iii) strong absorption of low-l partial waves.

Under these conditions we can replace the spherical harmonics in (8.2.3) by their asymptotic forms for large l_f. In this section we use the 'small-angle asymptotics'

$$Y_{l_f,-M}(\theta,0)\simeq\left(\frac{\lambda_f}{2\pi}\right)^{1/2}\left(\frac{\theta}{\sin\theta}\right)^{1/2}J_M(\lambda_f\theta), \tag{8.2.6}$$

where $\lambda_f=l_f+\frac{1}{2}$, which is valid for $0\leqslant\theta\lesssim\pi-|M|/\lambda_f$ and exact in the forward direction $\theta=0$. For $L\ll l_i$, l_f we can then also use the asymptotic form of the Clebsch–Gordan coefficients

$$\langle l_f L; -MM \mid l_i 0\rangle\simeq d^L_{MK}(\tfrac{1}{2}\pi) \tag{8.2.7}$$

in terms of reduced rotation matrix clements, where

$$K=l_i-l_f. \tag{8.2.8}$$

With the definitions

$$\begin{aligned} R^L_{l_f,l_f+K}(k_f,k_i)&\equiv R_{LK}(\lambda_f)\\ \sigma_{l_{f,i}}(k_{f,i})&\equiv\sigma^{(f,i)}(\lambda_{f,i}) \end{aligned} \tag{8.2.9}$$

and

$$b^L_{KM}=\mathrm{i}^K d^L_{0K}(\tfrac{1}{2}\pi)d^L_{MK}(\tfrac{1}{2}\pi), \tag{8.2.10}$$

we replace the summation over l_f in (8.2.3) by the Poisson series of integrals over λ_f with the result

$$\begin{aligned} \beta_{LM}(\theta)=\frac{\mathrm{i}^{-L}}{\pi^{1/2}}\left(\frac{\theta}{\sin\theta}\right)^{1/2}&\sum_K b^L_{KM}\sum_{mc-\infty}^{\infty}\exp(-\mathrm{i}m\pi)\\ &\times\int_0^\infty d\lambda_f\lambda_f R_{LK}(\lambda_f)\exp[\mathrm{i}\{\sigma^{(f)}(\lambda_f)+\sigma^{(i)}(\lambda_f+K)\}]\\ &\times\exp(\mathrm{i}2m\pi\lambda_f)J_M(\lambda_f\theta). \end{aligned} \tag{8.2.11}$$

Now we take the crucial step of making a suitable approximation for the functions $R_{LK}(\lambda_f)$. First we assume that the nuclear part of the distorting interaction in the initial and final channels is contained in these functions through the complex nuclear phases only, so that we can write

$$R_{LK}(\lambda_f)\approx\{S^{(N)}_{l_f}(k_f)\}^{1/2}\mathcal{R}^L_{l_f l_i}(k_f,k_i)\{S^{(N)}_{l_i}(k_i)\}^{1/2}, \tag{8.2.12}$$

where $S^{(N)}_l(k)$ is the nuclear part of the total elastic S-matrix,

$$\begin{aligned} S_l(k)=S^{(N)}_l(k)\exp\{\mathrm{i}2\sigma_l(k)\}&=\eta_l(k)\exp[\mathrm{i}2\{\delta^{(N)}_l(k)+\sigma_l(k)\}]\\ &=\eta_l(k)\exp\{\mathrm{i}2\delta_l(k)\}, \end{aligned} \tag{8.2.13}$$

and

$$\mathcal{R}^L_{l_f l_i}(k_f,k_i)=\frac{4\pi}{k_i k_f'}\int_0^\infty \mathrm{d}r f_{l_f}(k_f'r)F^{(\kappa)}_L(r)f_{l_i}(k_i r), \tag{8.2.14}$$

with $k_f' = \zeta k_f$, are the transfer radial integrals with *Coulomb* radial wave functions $f_l(kr)$ only. Approximations similar to (8.2.12), which is known as the 'Sopkovich prescription' (Sopkovich 1962 [8.29]; Gottfried and Jackson 1964 [8.30]), are usually made in semiclassical approaches (Broglia and Winther 1972 [8.17]; Landowne *et al.* 1976 [8.18]; Hasan and Brink 1978 [8.20]); the same device was used in our description of inelastic heavy-ion scattering (Frahn 1976 [8.27]) to approximate the Coulomb excitation part of the inelastic radial integrals.

Secondly, we replace the square root of the product of the nuclear S-matrices by an average S-matrix which can be regarded as the 'geometrical mean' of $S_{l_i}^{(N)}$ and $S_{l_f}^{(N)}$. If it is assumed that $S_{l_{i,f}}(k_{i,f})$ can be interpolated by smooth functions $S(\lambda_{i,f})$ of the continuous variables $\lambda_{i,f} = l_{i,f} + \frac{1}{2}$, characterized by critical angular momentum parameters $\Lambda_{i,f} = l_{i,f}^{(0)} + \frac{1}{2}$ and l-window parameters $\Delta_{i,f}$ on which they depend through the combinations $\mu_{i,f} = (\lambda_{i,f} - \Lambda_{i,f})/\Delta_{i,f}$, we can show (Appendix 1) that

$$\{S_{l_f}^{(N)}(k_f)S_{l_i}^{(N)}(k_i)\}^{1/2} \rightarrow \{S_N^{(f)}(\lambda_f)S_N^{(i)}(\lambda_i)\}^{1/2} \approx S_N^{(K)}(\bar{\lambda}) \quad (8.2.15)$$

where

$$S_N^{(K)}(\bar{\lambda}) = \eta_K(\bar{\lambda})\exp\{i2\delta_N^{(K)}(\bar{\lambda})\}, \qquad \bar{\lambda} = \tfrac{1}{2}(\lambda_i + \lambda_f), \quad (8.2.16)$$

with

$$2\delta_N^{(K)}(\bar{\lambda}) = \delta_N^{(i)}(\lambda_i) + \delta_N^{(f)}(\lambda_f) = \delta_N^{(i)}(\bar{\lambda} + \tfrac{1}{2}K) + \delta_N^{(f)}(\bar{\lambda} - \tfrac{1}{2}K) \quad (8.2.17)$$

and

$$\eta_K(\bar{\lambda}) \equiv \hat{\eta}(\bar{\mu}_K), \qquad \bar{\mu}_K = \frac{\bar{\lambda} - \Lambda_K}{\Delta}. \quad (8.2.18)$$

Here the critical angular momentum of the 'mean' reflection function is given by

$$\begin{aligned} \Lambda_K &= \bar{\Lambda} + \Delta_K \\ \Delta_K &= \Delta \ln\left[\frac{1}{2}\left\{\exp\left(\frac{K - K_0}{2\Delta_f}\right) + \exp\left(-\frac{K - K_0}{2\Delta_i}\right)\right\}\right] \end{aligned} \quad (8.2.19)$$

where

$$\bar{\Lambda} = \tfrac{1}{2}(\Lambda_i + \Lambda_f), \qquad K_0 = \Lambda_i - \Lambda_f \quad (8.2.20)$$

and Δ is the geometrical mean of the l-window parameters of the initial- and final-channel S-matrices:

$$\Delta = \frac{2\Delta_i\Delta_f}{\Delta_i + \Delta_f}. \quad (8.2.21)$$

It will be seen later on that the K and K_0 dependences of the mean S-matrix described by eqns (8.2.18)–(8.2.20) determine the magnetic substate population, the l-matching, and most of the Q-window effects in the transfer cross-section.

Thirdly, although the Coulomb radial integrals (8.2.14) can be expressed in terms of generalized hypergeometric functions of two variables known as Appell functions (Trautmann and Alder 1970 [8.31]; Alder and Trautmann 1971 [8.32]; Alder *et al.* 1972 [8.33]), for practical purposes we use an approximation obtained by replacing the Coulomb radial wave functions $\mathscr{f}_l(kr)$ by their WKB forms. Then we can write

$$\mathscr{R}^L_{l_f l_i}(k_f, k_i) = \frac{2\pi}{k_i k_f' \kappa} I^{(\kappa)}_{L-K}(\alpha, \xi), \tag{8.2.22}$$

where the $I^{(\kappa)}_{L-K}$ are functions of the mean asymptotic Rutherford deflection function α and the adiabaticity parameter ξ:

$$\alpha = \bar{\Theta}_R(\bar{\lambda}) = 2\arctan(n/\bar{\lambda}), \qquad \xi = n_f - n_i, \tag{8.2.23}$$

where

$$n = \tfrac{1}{2}(n_i + n_f) \tag{8.2.24}$$

$$n_i = \frac{\mu_i Z_{A_i} Z_{B_i} e^2}{\hbar^2 k_i} \qquad n_f = \frac{\mu_f Z_{A_f} Z_{B_f} e^2}{\hbar^2 k_f}. \tag{8.2.25}$$

The main properties of the functions $I^{(\kappa)}_{L-K}$ and some further approximations are given in Appendix 2.

Now we can write our approximation for the functions $R_{LK}(\lambda_f)$ as

$$R_{LK}(\lambda_f) = \frac{2\pi}{k_i k_f' \kappa} I^{(\kappa)}_{L-K}(\alpha, \xi)\eta_K(\lambda_f + \tfrac{1}{2}K)\exp\{i2\delta^{(K)}_N(\lambda_f + \tfrac{1}{2}K)\}, \tag{8.2.26}$$

and the amplitude (8.2.11) becomes

$$\begin{aligned}\beta_{LM}(\theta) = {} & i^{-L}\frac{2\pi^{1/2}}{k_i k_f' \kappa}\left(\frac{\theta}{\sin\theta}\right)^{1/2} \sum_K b^L_{KM} \sum_{m=-\infty}^{\infty} \exp(-im\pi) \\ & \times \int_0^\infty d\lambda_f \lambda_f I^{(\kappa)}_{L-K}(\alpha, \xi)\eta_K(\lambda_f + \tfrac{1}{2}K) \\ & \times \exp[i\{2\delta_K(\lambda_f + \tfrac{1}{2}K) + 2m\pi\lambda_f\}]J_M(\lambda_f\theta),\end{aligned} \tag{8.2.27}$$

where we have defined the mean total phase shift function

$$\begin{aligned}2\delta_K(\bar{\lambda}) &= \delta^{(i)}_N(\lambda_i) + \sigma^{(i)}(\lambda_i) + \delta^{(f)}_N(\lambda_f) + \sigma^{(f)}(\lambda_f) \\ &\equiv 2\delta^{(K)}_N(\lambda_f + \tfrac{1}{2}K) + 2\sigma^{(K)}(\lambda_f + \tfrac{1}{2}K).\end{aligned} \tag{8.2.28}$$

To evaluate the integrals over λ_f in (8.2.27) we first note that the nonoscillatory part of the integrands are localized in l-space. This is because the reflection functions η_K for heavy ions have a strong-absorption profile with very small values for low-l partial waves and because the functions $I^{(\kappa)}_{L-K}$ fall off exponentially at large l with a rate determined by the binding parameter κ. Thus the 'transfer partial-wave amplitude functions'

$$c_{LK}(\bar{\lambda}) = I^{(\kappa)}_{L-K}(\alpha, \xi)\eta_K(\bar{\lambda}), \tag{8.2.29}$$

whose general shape is sketched in Fig. 8.1, have maxima at $\lambda = \bar{\Lambda}^T_K$. We note that, in contrast with the simple parametric forms assumed by Frahn and Venter (1964 [8.1]), Strutinsky (1964 [8.2]), Chasman *et al.* (1973 [8.21]), and Kahana *et al.* (1974 [8.23]), the functions $c_{LK}(\bar{\lambda})$ derived from distorted-wave theory are *essentially asymmetric* about their maximum positions, the asymmetry being determined by the difference between the binding parameter κ/k and the mean elastic window parameter Δ, i.e. by the difference between the range R_B of the bound-state wave function of the transferred cluster and the mean surface diffuseness $\bar{a}$ semiclassically associated with Δ.

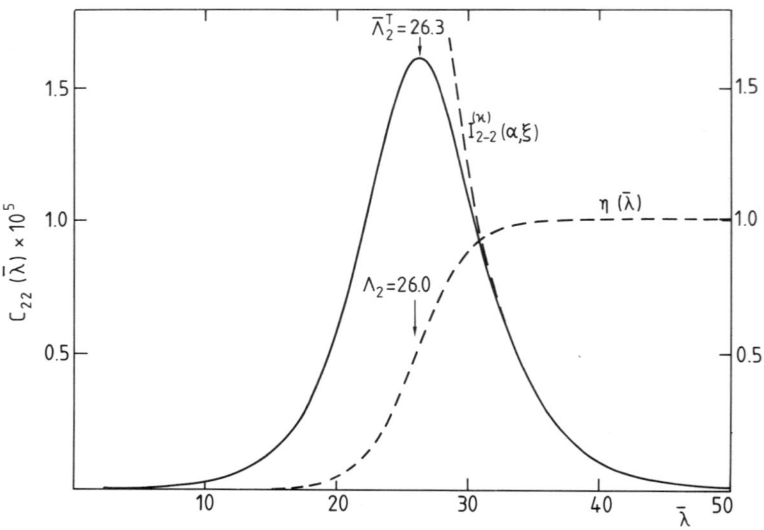

FIG. 8.1. Typical shape of the 'transfer partial-wave amplitude function' $c_{LK}(\bar{\lambda})$ defined by eqn (8.2.29) as a function of the mean orbital angular momentum variable $\bar{\lambda} = \frac{1}{2}(\lambda_i + \lambda_f)$, for $L = K = 2$ (——). The example shown pertains to the reaction $^{48}Ca(^{16}O, ^{12}C)^{50}Ti$ at $E_{lab} = 56$ MeV. The functions $I^{(\kappa)}_{L-K}(\alpha, \xi)$ and $\eta_K(\bar{\lambda})$ of which $c_{LK}(\bar{\lambda})$ is the product are also shown; $I^{(\kappa)}_{2-2}(\alpha, \xi)$ with $\xi = -2.79$, $\kappa = 1.172$, and $\gamma = 0.2667$ (– – –) and $\eta_2(\bar{\lambda})$ with $\Lambda_2 = 26.0$ and $\Delta = 2$ (- - -). The different scale on right-hand side should be noted.

From Appendix 2 it is seen that we can write

$$I^{(\kappa)}_{L-K}(\alpha,\xi) \equiv \tilde{I}^{(\kappa)}_{L-K}(\alpha,\xi)\exp(-\gamma\bar{\lambda})$$
$$= \tilde{I}^{(\kappa)}_{L-K}(\alpha,\xi)\exp(-\tfrac{1}{2}\gamma K)\exp(-\gamma\lambda_f), \qquad (8.2.30)$$

where $\tilde{I}^{(\kappa)}_{L-K}(\alpha,\xi)$ is a slowly varying function of $\bar{\lambda}$, and where

$$\gamma = (\rho^2 + \xi'^2)^{1/2}/n \qquad (8.2.31)$$

with

$$\rho = \frac{\kappa}{k}\,n, \qquad k = \tfrac{1}{2}(k_i + k_f), \qquad \xi' = \frac{k_i - k'_f}{k}\,n. \qquad (8.2.32)$$

In terms of λ_f, the maximum of $c_K(\bar{\lambda}) = c_K(\lambda_f + \frac{1}{2}K)$ is at $\lambda_f = \Lambda^T_K = \bar{\Lambda}^T_K - \frac{1}{2}K$. Its value can be calculated approximately by determining the maximum of

$$\tau_K(\lambda_f) = \eta_K(\lambda_f + \tfrac{1}{2}K)\exp(-\gamma\lambda_f), \qquad (8.2.33)$$

which yields

$$\Lambda^T_K = \Lambda_K - \tfrac{1}{2}K + \delta_T, \qquad \bar{\Lambda}^T_K = \Lambda_K + \delta_T, \qquad (8.2.34)$$

where Λ_K is given by (8.2.19) and δ_T is obtained by solving the equation

$$\frac{\hat{D}(\delta_T/\Delta)}{\hat{\eta}(\delta_T/\Delta)} = \gamma\Delta \qquad (8.2.35)$$

with the definition $\hat{D}(\mu) = \mathrm{d}\hat{\eta}(\mu)/\mathrm{d}\mu$. As an example, for the convenient parametric form $\hat{\eta}(\mu) = (1 + \mathrm{e}^{-\mu})^{-1}$ we obtain

$$\delta_T = \Delta \ln\left(\frac{1}{\gamma\Delta} - 1\right) \qquad (8.2.36)$$

with the condition $\gamma\Delta < 1$. Notice that, quite generally, the shift δ_T is independent of K.

We now write

$$c_{LK}(\bar{\lambda}) = \tilde{c}_{LK}(\lambda_f)\tau_K(\lambda_f)$$
$$= \tilde{I}^{(\kappa)}_{L-K}(\alpha,\xi)\exp(-\tfrac{1}{2}\gamma K)\eta_K(\lambda_f + \tfrac{1}{2}K)\exp(-\gamma\lambda_f) \qquad (8.2.37)$$

and expand the phase functions in (8.2.27) linearly about $\lambda_f = \Lambda^T_K$:

$$2\delta_K(\lambda_f + \tfrac{1}{2}K) \approx 2\delta_K(\Lambda^T_K + \tfrac{1}{2}K) + (\lambda_f - \Lambda^T_K)\theta^T_K$$
$$= 2\delta_K(\bar{\Lambda}^T_K) + (\lambda_f - \Lambda^T_K)\theta^T_K, \qquad (8.2.38)$$

where

$$\theta^T_K = \Theta(\bar{\Lambda}^T_K) = \left\{\frac{2\delta_K(\bar{\lambda})}{\mathrm{d}\bar{\lambda}}\right\}_{\bar{\lambda} = \bar{\Lambda}_K} \qquad (8.2.39)$$

are the critical angles associated with the characteristic angular momenta $\bar{\Lambda}_K^T$. After taking the slowly varying functions $\tilde{c}_{LK}(\lambda_f)$ from under the integrals at the points $\lambda_f = \Lambda_K^T$, we obtain from (8.2.27)

$$\beta_{LM}(\theta) = i^{-L} \frac{2\pi^{1/2}}{k_i k_f' \kappa} \left(\frac{\theta}{\sin\theta}\right)^{1/2} \sum_K b_{KM}^L \tilde{c}_{LK}(\Lambda_K^T) \exp\{i2\delta_K(\bar{\Lambda}_K^T)\}$$
$$\times \sum_{m=-\infty}^{\infty} \exp\{i2m\pi(\Lambda_K^T - \tfrac{1}{2})\} T_{KM}(\theta_K^T + 2m\pi), \tag{8.2.40}$$

where

$$T_{KM}(x) = \int_0^\infty d\lambda\, \lambda \tau_K(\lambda) \exp\{i(\lambda - \Lambda_K^T)x\} J_M(\lambda\theta). \tag{8.2.41}$$

These integrals are the type of combined Fourier and Bessel transforms that were evaluated in the Appendix of Frahn (1980 [8.34]) with the result

$$T_{KM}(x) = \frac{\Lambda_K^T}{2} [H_K^{(T)}(x-\theta)\{J_M(\Lambda_K^T\theta) - iJ_{M+1}(\Lambda_K^T\theta)\}$$
$$+ H_K^{(T)}(x+\theta)\{J_M(\Lambda_K^T\theta) + iJ_{M+1}(\Lambda_K^T\theta)\}], \tag{8.2.42}$$

where $H_K^{(T)}(z)$ is the Fourier transform of $\tau_K(\lambda)$:

$$H_K^{(T)}(z) = \int_{-\infty}^\infty d\lambda\, \tau_K(\lambda) \exp\{i(\lambda - \Lambda_K^T)z\}. \tag{8.2.43}$$

By using eqn (8.2.33) we obtain

$$H_K^{(T)}(z) = \int_{-\infty}^\infty d\lambda\, \eta_K(\lambda + \tfrac{1}{2}K) \exp(-\gamma\lambda) \exp\{i(\lambda - \Lambda_K^T)z\}$$
$$= \exp(-\gamma\Lambda_K^T) \int_{-\infty}^\infty d\bar{\lambda}\, \eta_K(\bar{\lambda}) \exp\{i(\bar{\lambda} - \bar{\Lambda}_K^T)(z + i\gamma)\}$$
$$= \exp(-\gamma\Lambda_K^T) \exp\{i(\Lambda_K - \bar{\Lambda}_K^T)(z + i\gamma)\}$$
$$\times \int_{-\infty}^\infty d\bar{\lambda}\, \eta_K(\bar{\lambda}) \exp\{i(\bar{\lambda} - \Lambda_K)(z + i\gamma)\}$$
$$= i \exp(\tfrac{1}{2}\gamma K) \exp(-\gamma\Lambda_K) \exp(-i\delta_T z) \frac{F\{\Delta(z + i\gamma)\}}{z + i\gamma}, \tag{8.2.44}$$

where

$$F(\Delta x) = \int_{-\infty}^\infty d\lambda \frac{d\eta_K(\lambda)}{d\lambda} \exp\{i(\lambda - \Lambda_K)x\} \tag{8.2.45}$$

is the Fourier transform of the derivative of the mean reflection function.

Noting from (8.2.38) that

$$2\delta_K(\bar{\Lambda}_K^T) - \delta_T\theta_K^T \approx 2\delta_K(\Lambda_K), \tag{8.2.46}$$

and from (8.2.30) and (8.2.37) that

$$\tilde{c}_{LK}(\Lambda_K^T)\exp(\tfrac{1}{2}\gamma K)\exp(-\gamma\Lambda_K) = I_{L-K}^{(\kappa)}(\theta_{R,K}^T, \xi)\exp(\gamma\delta_T), \tag{8.2.47}$$

where

$$\theta_{R,K}^T = 2\arctan(n/\bar{\Lambda}_K^T), \tag{8.2.48}$$

we can write the final result as

$$\begin{aligned}\beta_{LM}(\theta) = \mathrm{i}^{1-L}\frac{\pi^{1/2}}{k_i k_f' \kappa}\left(\frac{\theta}{\sin\theta}\right)^{1/2} & \sum_K b_{KM}^L \Lambda_K^T I_{L-K}^{(\kappa)}(\theta_{R,K}^T, \xi)\\ & \times \exp(\gamma\delta_T)\exp\{\mathrm{i}2\delta_K(\Lambda_K)\}\\ & \times [\mathscr{H}_K^{(SA)}(\theta)\{J_M(\Lambda_K^T\theta) - \mathrm{i}J_{M+1}(\Lambda_K^T\theta)\}\\ & + \mathscr{H}_K^{(SA)}(-\theta)\{J_M(\Lambda_K^T\theta) + \mathrm{i}J_{M+1}(\Lambda_K^T\theta)\}],\end{aligned} \tag{8.2.49}$$

where

$$\begin{aligned}\mathscr{H}_K^{(SA)}(\pm\theta) = \exp(\pm \mathrm{i}\delta_T\theta) & \sum_{m=-\infty}^{\infty} \exp[\mathrm{i}2m\pi\{\Lambda_K - \tfrac{1}{2}(K+1)\}]\\ & \times \frac{F\{\Delta(\theta_K^T + 2m\pi + \mathrm{i}\gamma \mp \theta)\}}{\theta_K^T + 2m\pi + \mathrm{i}\gamma \mp \theta}.\end{aligned} \tag{8.2.50}$$

Close to the forward direction we have approximately

$$\begin{aligned}\beta_{LM}(\theta) \approx \mathrm{i}^{1-L}\frac{2\pi^{1/2}}{k_i k_f' \kappa} & \sum_K b_{KM}^L \Lambda_K^T I_{L-K}^{(\kappa)}(\theta_{R,K}^T, \xi)\\ & \times \exp(\gamma\delta_T)\exp\{\mathrm{i}2\delta_K(\Lambda_K)\}\mathscr{H}_K^{(SA)}(0)J_M(\Lambda_K^T\theta), \qquad \theta \approx 0,\end{aligned} \tag{8.2.51}$$

so that the near-forward cross-section behaves like

$$\sigma_{LM}^{(T)}(\theta) \propto \{J_M(\Lambda_k^T\theta)\}^2 \propto \theta^{2|M|}, \qquad \theta \approx 0. \tag{8.2.52}$$

8.2.2 *Forward-angle excitation function*

In the forward direction $\theta = 0$ only the $M = 0$ substate contributes, and we find

$$\begin{aligned}\beta_{LM}(0) &= \beta_{L0}(0)\delta_{M0}\\ \beta_{L0}(0) &= \frac{2\pi^{1/2}}{k_i k_f' \kappa}\sum_K \mathrm{i}^{K-L+1}\{d_{0K}^L(\tfrac{1}{2}\pi)\}^2 \Lambda_K^T I_{L-K}^{(\kappa)}(\theta_{R,K}^T, \xi)\\ &\qquad \times \exp(\gamma\delta_T)\exp\{\mathrm{i}2\delta_K(\Lambda_K)\}\mathscr{H}_K^{(SA)}(0),\end{aligned} \tag{8.2.53}$$

where

$$\mathscr{H}_K^{(\mathrm{SA})}(0)=\sum_{m=-\infty}^{\infty}\exp[\mathrm{i}2m\pi\{\Lambda_K-\tfrac{1}{2}(K+1)\}]\frac{F\{\Delta(\theta_K^T+2m\pi+\mathrm{i}\gamma)\}}{\theta_K^T+2m\pi+\mathrm{i}\gamma}. \quad (8.2.54)$$

To discuss the energy dependence we consider $L=0$ transfer only. Then the forward-angle cross-section becomes

$$\sigma_0^{(T)}(0)=A\frac{\mu_\mathrm{i}\mu_\mathrm{f}}{\pi\hbar^4\zeta^2k_\mathrm{i}^3k_\mathrm{f}\kappa^2}(\Lambda_0^T)^2\{I_{00}^{(\kappa)}(\theta_R^T,\xi)\}^2 \times\exp(2\gamma\delta_T)\,|\mathscr{H}_0^{(\mathrm{SA})}(0)|^2. \quad (8.2.55)$$

From the expression for $I_{00}^{(\kappa)}$ given in Appendix 2 it is seen that for $\Lambda_0^T\gg n$,

$$I_{00}^{(\kappa)}(\theta_R^T,\xi)\approx\left(\frac{\pi}{2\gamma\Lambda_0^T}\right)^{1/2}\exp\left\{-\left(\rho+\xi\arctan\frac{\xi'}{\rho}\right)\right\}\exp(-\gamma\Lambda_0^T)$$
$$\approx\left(\frac{\pi k}{2\kappa\Lambda_0^T}\right)^{1/2}\exp(-\gamma\Lambda_0^T), \quad (8.2.56)$$

so that

$$\sigma_0^{(T)}(0)\approx A\frac{\mu_\mathrm{i}\mu_\mathrm{f}}{2\hbar^4\zeta^2\kappa^2k^3}\Lambda_0^T\exp\left(-\frac{2\kappa\Lambda_0}{k}\right)|\mathscr{H}_0^{(\mathrm{SA})}(0)|^2 \quad (8.2.57)$$

with

$$\Lambda_0=\bar{\Lambda}+\Delta\ln\left\{\cosh\left(\frac{K_\mathrm{i}}{2\Delta}\right)\right\}, \quad (8.2.58)$$

where for simplicity we have assumed $\Delta_\mathrm{i}=\Delta_\mathrm{f}=\Delta$. Since Λ_0/k depends only weakly on energy well above the Coulomb barrier, the main energy dependence is described by $|\mathscr{H}_0^{(\mathrm{SA})}(0)/k|^2$.

Normally only the $m=0$ term gives a significant contribution to the Poisson sum (8.2.54) so that

$$\mathscr{H}_0^{(\mathrm{SA})}(0)=\frac{F\{\Delta(\theta_0^T+\mathrm{i}\gamma)\}}{\theta_0^T+\mathrm{i}\gamma}. \quad (8.2.59)$$

As a typical example, for the parametric form $\hat{\eta}(\mu)=(1+\mathrm{e}^{-\mu})^{-1}$ we have $F(z)=\pi z/\sinh(\pi z)$, so that

$$\left|\frac{\mathscr{H}_0^{(\mathrm{SA})}(0)}{k}\right|^2=\frac{(\pi\Delta/k)^2}{\{\sinh(\pi\Delta\theta_0^T)\}^2+\{\sin(\pi\Delta\gamma)\}^2}. \quad (8.2.60)$$

Since $\Delta\gamma\approx\kappa(\Delta/k)\approx$constant, the main energy dependence comes from $\Delta\theta_0^T\approx n(\Delta/\Lambda_0^T)$ and is thus chiefly determined by the energy variation of the Sommerfeld parameter n, a feature similar to that found for elastic and inelastic heavy-ion scattering (Frahn 1975 [8.25]; Frahn 1975 [8.26]; Frahn 1976 [8.27]). In general, the structure of eqns (8.2.53) and (8.2.54) shows that as long as only a single ($m=0$) Poisson term is significant, the forward-angle transfer excitation function is smooth or at least is not strongly oscillatory.

8.3 Large-angle representation

8.3.1 *Angular distribution*

To evaluate the transfer amplitude for large angles θ we use instead of (8.2.6) the large-angle asymptotic form of the spherical harmonics

$$Y_{l_f,-M}(\theta,0) \simeq (-)^{l_f+M}\left(\frac{\lambda_f}{2\pi}\right)^{1/2}\left(\frac{\vartheta}{\sin\vartheta}\right)^{1/2} J_M(\lambda_f\vartheta), \tag{8.3.1}$$

where $\vartheta=\pi-\theta$, which is valid for $0\leqslant\vartheta<\pi-|M|/\lambda_f$ and is exact in the backward direction $\theta=\pi$. Then the Poisson representation becomes

$$\begin{aligned}\beta_{LM}(\theta) = {} & \frac{i^{-L}}{\pi^{1/2}}(-)^M\left(\frac{\vartheta}{\sin\vartheta}\right)^{1/2}\sum_K b^L_{KM}\sum_{m=-\infty}^{\infty}\exp\{-i(m+\tfrac{1}{2})\pi\} \\ & = \int_0^\infty d\lambda_f\lambda_f R_{LK}(\lambda_f)\exp[i\{\sigma^{(f)}(\lambda_f)+\sigma^{(i)}(\lambda_f+K)\}] \\ & \times\exp\{i(2m+1)\pi\lambda_f\}J_M(\lambda_f\vartheta).\end{aligned} \tag{8.3.2}$$

With the same approximations as used in the previous section we obtain

$$\begin{aligned}\beta_{LM}(\theta) = {} & i^{1-L}(-)^M\frac{\pi^{1/2}}{k_ik'_{fK}}\left(\frac{\vartheta}{\sin\vartheta}\right)^{1/2}\sum_K b^L_{KM}\Lambda^T_K I^{(\kappa)}_{L-K}(\theta^T_{R,K},\xi) \\ & \times\exp(\gamma\delta_T)\exp\{i2\delta_K(\Lambda_K)\} \\ & = \mathscr{H}^{(LA)}_K(\vartheta)\{J_M(\Lambda^T_K\vartheta)-iJ_{M+1}(\Lambda^T_K\vartheta)\} \\ & + \mathscr{H}^{(LA)}_K(-\vartheta)\{J_M(\Lambda^T_K\vartheta)+iJ_{M+1}(\Lambda^T_K\vartheta)\},\end{aligned} \tag{8.3.3}$$

where

$$\begin{aligned}\mathscr{H}^{(LA)}_K(\pm\vartheta) = {} & \exp(\pm i\delta_T\vartheta)\sum_{m=-\infty}^{\infty}\exp[i(2m+1)\pi\{\Lambda_K-\tfrac{1}{2}(K+1)\}] \\ & \times\frac{F\{\Delta(\theta^T_K+(2m+1)\pi+i\gamma\mp\vartheta)\}}{\theta^T_K+(2m+1)\pi+i\gamma\mp\vartheta}.\end{aligned} \tag{8.3.4}$$

For later use we note the following relation which is obtained by comparing (8.3.4) with (8.2.50):

$$\mathscr{H}^{(LA)}_K(\pm\vartheta) = \exp\{\pm i\pi(\Lambda^T_K-\tfrac{1}{2})\}\mathscr{H}^{(SA)}_K(\mp\theta). \tag{8.3.5}$$

Close to the backward direction we have approximately

$$\beta_{LM}(\theta)\approx \mathrm{i}^{1-L}(-)^M\frac{2\pi^{1/2}}{k_\mathrm{i}k_\mathrm{f}'\kappa}\sum_K b_{KM}^L\Lambda_K^T I_{L-K}^{(\kappa)}(\theta_{\mathrm{R,K}}^T,\xi)$$
$$\times\exp(\gamma\delta_T)\exp\{\mathrm{i}2\delta_K(\Lambda_K)\}$$
$$\times\mathcal{H}_K^{(\mathrm{LA})}(0)J_M\{\Lambda_K^T(\pi-\theta)\},\qquad \theta\approx\pi, \tag{8.3.6}$$

so that the near-backward cross-section behaves like

$$\sigma_{LM}^{(T)}(\theta)\propto[J_M\{\Lambda_K^T(\pi-\theta)\}]^2\propto(\pi-\theta)^{2|M|},\qquad \theta\approx\pi. \tag{8.3.7}$$

8.3.2 *Backward-angle excitation function*

In the backward direction $\theta=\pi$ only the $M=0$ substate contributes, and (8.3.6) reduces to

$$\beta_{LM}(\pi)=\beta_{L0}(\pi)\delta_{M0},$$
$$\beta_{L0}(\pi)=\frac{2\pi^{1/2}}{k_\mathrm{i}k_\mathrm{f}'\kappa}\sum_K \mathrm{i}^{K-L+1}\{d_{0K}^L(\tfrac{1}{2}\pi)\}^2\Lambda_K^T I_{L-K}^{(\kappa)}(\theta_{\mathrm{R,K}}^T,\xi) \tag{8.3.8}$$
$$\times\exp(\gamma\delta_T)\exp\{\mathrm{i}2\delta_K(\Lambda_K)\}\mathcal{H}_K^{(\mathrm{LA})}(0).$$

Comparison with (8.2.53) and use of relation (8.3.5) shows that the relative magnitude of $\beta_{L0}(\pi)$ and $\beta_{L0}(0)$ is determined by the ratio $\mathcal{H}_K^{(\mathrm{SA})}(-\pi)/\mathcal{H}_K^{(\mathrm{SA})}(0)$. If, as is normally the case, the $m=0$ term dominates in the Poisson sum, this becomes

$$\left|\frac{\mathcal{H}_K^{(\mathrm{SA})}(-\pi)}{\mathcal{H}_K^{(\mathrm{SA})}(0)}\right|=\left\{\frac{(\theta_K^T)^2+\gamma^2}{(\theta_K^T+\pi)^2+\gamma^2}\right\}^{1/2}\left|\frac{F\{\Delta(\theta_K^T+\pi+\mathrm{i}\gamma)\}}{F\{\Delta(\theta_K^T+\mathrm{i}\gamma)\}}\right|\ll 1. \tag{8.3.9}$$

For example, with $F(z)=\pi z/\sinh(\pi z)$ we have for the $L=0$ backward-to-forward cross-section ratio

$$\frac{\sigma_0^{(T)}(\pi)}{\sigma_0^{(T)}(0)}=\left|\frac{\mathcal{H}_0^{(\mathrm{SA})}(-\pi)}{\mathcal{H}_0^{(\mathrm{SA})}(0)}\right|^2=\frac{4\exp(-2\pi^2\Delta)}{|1+\coth\{\pi\Delta(\theta_0^T+\mathrm{i}\gamma)\}|^2}. \tag{8.3.10}$$

This shows that the backward-angle transfer excitation function has an additional energy dependence: the ratio (8.3.10) decreases rapidly with increasing energy since $\Delta\propto E^{1/2}$.

8.3.3 *Enhancement effects*

In our discussion so far we have assumed that the elastic S-matrices in the initial and final channels have 'normal strong-absorption profile', in the terminology of Frahn (1980 [8.34]). As described there backward-angle enhancement can occur for elastic and inelastic scattering if $\eta_K(\bar{\lambda})$ contains rapidly varying deviations from the normal $\bar{\lambda}$ profile. If we write

$$\eta_K(\bar{\lambda}) = \bar{\eta}_K(\bar{\lambda}) + \tilde{\eta}_K(\bar{\lambda}), \tag{8.3.11}$$

then the transfer amplitude has an additional contribution $\tilde{\beta}(\theta)$ of a form similar to (8.3.3) but with the functions $\mathscr{H}_K^{(\text{LA})}(\pm\vartheta)$ replaced by

$$\tilde{\mathscr{H}}_K^{(\text{LA})}(\pm\vartheta) = -\mathrm{i}\exp(\pm\mathrm{i}\delta_T\vartheta)\sum_{m=-\infty}^{\infty}\exp[\mathrm{i}(2m+1)\pi\{\Lambda_K - \tfrac{1}{2}(K+1)\}]$$
$$\times G\{\theta_K^T + (2m+1)\pi + \mathrm{i}\gamma \mp \vartheta\}, \tag{8.3.12}$$

where

$$G(x) = \int_{-\infty}^{\infty} \mathrm{d}\lambda\,\tilde{\eta}_K(\lambda)\exp\{\mathrm{i}(\lambda - \Lambda_K)x\} \tag{8.3.13}$$

and where we have assumed for simplicity that the deviation $\tilde{\eta}_K(\bar{\lambda})$ is centred at $\bar{\lambda} = \Lambda_K$. Then the condition for backward-angle enhancement is

$$\left|\frac{\tilde{\mathscr{H}}_K^{(\text{LA})}(0)}{\mathscr{H}_K^{(\text{LA})}(0)}\right|^2 \approx \left|\frac{\theta_K^T - \pi + \mathrm{i}\gamma}{F[\Delta(\theta_K^T - \pi + \mathrm{i}\gamma)]}G(\theta_K^T - \pi + \mathrm{i}\gamma)\right|^2 \gg 1 \tag{8.3.14}$$

if the $m = -1$ contributions to the Poisson sums in (8.3.12) and (8.3.4) are dominant. Thus, except for the effects of the transfer parameters γ and δ_T, the enhancement conditions are much the same as for the elastic and inelastic scattering described by Frahn (1980 [8.34]).

Interference between the $m = -1$ and $m = 0$ terms in (8.3.12) gives rise to oscillations in the backward-angle excitation function. Assuming $L = 0$ transfer for simplicity, we obtain

$$\tilde{\sigma}_0(\pi) = \tilde{\sigma}_0^{(0)}(\pi)\left|1 - \frac{G(\theta_0^T + \pi + \mathrm{i}\gamma)}{G(\theta_0^T - \pi + \mathrm{i}\gamma)}\exp(\mathrm{i}2\pi\Lambda_0)\right|^2 \tag{8.3.15}$$

which shows that, apart from more slowly varying modulations due to γ and δ_T the dominant oscillations are of period $\delta\Lambda_0 = 1$ as in the case of elastic and inelastic scattering. (Odd–even staggering effects in $\tilde{\eta}(\bar{\lambda})$ would cause an additional modulation of period $\delta\Lambda_0 = 2$.)

8.4 Intermediate angles

8.4.1 *Angular distribution*

In this section we consider the intermediate-angle region for which we can use the asymptotic expression

$$J_M(x) \pm \mathrm{i}J_{M+1}(x) \simeq \left(\frac{2}{\pi x}\right)^{1/2}\exp\{\pm\mathrm{i}(x - \tfrac{1}{4}\pi - \tfrac{1}{2}M\pi)\}. \tag{8.4.1}$$

Then (8.2.49) becomes

$$\beta_{LM}(\theta) = \frac{1}{(\sin\theta)^{1/2}} \sum_K B^L_{KM}[\mathcal{H}^{(SA)}_K(\theta)\exp\{-i(\Lambda^T_K\theta - \tfrac{1}{4}\pi)\}$$
$$+(-)^M \mathcal{H}^{(SA)}_K(-\theta)\exp\{i(\Lambda^T_K\theta - \tfrac{1}{4}\pi)\}], \tag{8.4.2}$$

where

$$B^L_{KM} = \frac{i^{M-L+1}}{k_i k_f' \kappa} b^L_{KM}(2\Lambda^T_K)^{1/2} I^{(\kappa)}_{L-K}(\theta^T_{R,K}, \xi)\exp(\gamma\delta_T)\exp\{i2\delta_K(\Lambda_K)\}. \tag{8.4.3}$$

Similarly, we obtain from (8.3.3)

$$\beta_{LM}(\theta) = \frac{1}{(\sin\theta)^{1/2}} \sum_K B^L_{KM}[\mathcal{H}^{(LA)}_K(-\vartheta)$$
$$\times \exp\{i\pi(\Lambda^T_K - \tfrac{1}{2})\}\exp\{-i(\Lambda^T_K - \tfrac{1}{4}\pi)\}$$
$$+(-)^M \mathcal{H}^{(LA)}_K(\vartheta)\exp\{-i\pi(\Lambda^T_K - \tfrac{1}{2})\}\exp\{i(\Lambda^T_K - \tfrac{1}{4}\pi)\}], \tag{8.4.4}$$

which is the same as (8.4.2) by virtue of the relation (8.3.5). Thus the expressions (8.2.49) and (8.3.3) cover the full angular range, with the common simplified formula (8.4.2) for the intermediate region.

It is now convenient to change coordinate axes by transformation to a frame with the z-axis in the direction of $\mathbf{k}_i \times \mathbf{k}_f$ and the x-axis in the direction of $\mathbf{k}_i$. This rotation is performed by

$$\hat{\beta}_{LM}(\theta) = (-)^M \sum_{M'} i^{M'} d^L_{M'M}(\tfrac{1}{2}\pi)\beta_{LM'}(\theta). \tag{8.4.5}$$

Then the summations over M' and K can be carried out using the symmetry and orthogonality relations of the rotation matrices:

$$\sum_{M'} d^L_{M'M}(\tfrac{1}{2}\pi) d^L_{M'K}(\tfrac{1}{2}\pi) = \delta_{MK} \tag{8.4.6}$$

$$\sum_{M'} (-)^{M'} d^L_{M'M}(\tfrac{1}{2}\pi) d^L_{M'K}(\tfrac{1}{2}\pi) = (-)^L \delta_{M,-K}. \tag{8.4.7}$$

The resulting amplitude can be written

$$\hat{\beta}_{LM}(\theta) = \beta^{(+)}_{LM}(\theta) + (-)^M \beta^{(-)}_{LM}(\theta), \qquad L+M \text{ even} \tag{8.4.8}$$

(on the understanding that M from now on refers to the normal of the scattering plane as quantization axis), where

$$\beta^{(\pm)}_{LM}(\theta) = y_{LM}(\theta) B^{(\pm)}_{LM} \mathcal{H}^{(\pm)}_M(\theta)\exp\{\mp i(\Lambda_{\mp M}\theta - \tfrac{1}{4}\pi)\} \tag{8.4.9}$$

with

$$y_{LM}(\theta) = \frac{d^L_{0M}(\tfrac{1}{2}\pi)\exp\{i\tfrac{1}{2}M(\pi-\theta)\}}{(4\pi\sin\theta)^{1/2}} = \frac{Y_{LM}(\tfrac{1}{2}\pi, 0)}{(2L+1)^{1/2}} \frac{\exp\{i\tfrac{1}{2}M(\pi-\theta)\}}{(\sin\theta)^{1/2}} \tag{8.4.10}$$

$$B^{(\pm)}_{LM} = \mathrm{i}^{1-L}\frac{(8\pi)^{1/2}}{k_\mathrm{i}k'_\mathrm{f}\kappa}(\Lambda^T_{\mp M})^{1/2} I^{(\kappa)}_{L,\pm M}(\theta^T_{R,\mp M}, \xi)$$
$$\times \exp(\gamma\delta_T)\exp\{\mathrm{i}2\delta_{\mp M}(\Lambda_{\mp M})\} \quad (8.4.11)$$

$$\mathscr{H}^{(\pm)}_M(\theta) = \sum_{m=-\infty}^{\infty} (-)^{m(M+1)}\exp(\mathrm{i}2m\pi\Lambda_{\mp M})$$
$$\frac{F\{\Delta(\theta^T_{\mp M} + 2m\pi + \mathrm{i}\gamma \mp \theta)\}}{\theta^T_{\mp M} + 2m\pi + \mathrm{i}\gamma \mp \theta}. \quad (8.4.12)$$

Thus the transfer cross-section for multipolarity (L, M) can be written as

$$\sigma^{(T)}_{LM}(\theta) = \bar{\sigma}^{(T)}_{LM}(\theta)[\mathrm{i} + |t_{LM}(\theta)|^2 + (-)^M 2\,\mathrm{Im}\{t_{LM}(\theta)\exp(\mathrm{i}2\bar{\Lambda}_M\theta)\}], \quad (8.4.13)$$

where

$$\bar{\sigma}^{(T)}_{LM}(\theta) = A\frac{\mu_\mathrm{i}\mu_\mathrm{f}}{2\pi^2\hbar^4\zeta^2k^3_\mathrm{i}k_\mathrm{f}\kappa^2}\{d^L_{0M}(\tfrac{1}{2}\pi)\}^2\Lambda^T_{-M}$$
$$\times\{I^{(\kappa)}_{LM}(\theta^T_{R,-M}, \xi)\}^2\exp(2\gamma\delta_T)\frac{|\mathscr{H}^{(+)}_M(\theta)|^2}{\sin\theta}, \quad (8.4.14)$$

$$t_{LM}(\theta) = \left(\frac{\Lambda^T_M}{\Lambda^T_{-M}}\right)^{1/2}\frac{I^{(\kappa)}_{L-M}(\theta^T_{R,M}, \xi)}{I^{(\kappa)}_{LM}(\theta^T_{R,-M}, \xi)}\exp\{\mathrm{i}(\Delta_M - \Delta_{-M})\bar{\theta}\}\frac{\mathscr{H}^{(-)}_M(\theta)}{\mathscr{H}^{(+)}_M(\theta)}, \quad (8.4.15)$$

and

$$\bar{\Lambda}_M = \tfrac{1}{2}(\Lambda_M + \Lambda_{-M}) = \bar{\Lambda} + \tfrac{1}{2}(\Delta_M + \Delta_{-M}). \quad (8.4.16)$$

In (8.4.15) we have used

$$2\{\delta_M(\Lambda_M) - \delta_{-M}(\Lambda_{-M})\} \approx (\Delta_M - \Delta_{-M})\bar{\theta} \quad (8.4.17)$$

where $\bar{\theta} = \Theta(\bar{\Lambda})$.

Equation (8.4.14) shows that the transfer angular distribution has a smooth ('bell-shaped') part determined by

$$\frac{1}{\sin\theta}\{|\mathscr{H}^{(+)}_M(\theta)|^2 + |\mathscr{H}^{(-)}_M(\theta)|^2\} \quad (8.4.18)$$

and an oscillatory part described by the term in the curly brackets in eqn (8.4.13). Under normal conditions only the $m = 0$ term in the Poisson sum (8.4.12) is significant, so that

$$\mathscr{H}^{(+)}_M(\theta) \approx \frac{F\{\Delta(\theta^T_{-M} - \theta + \mathrm{i}\gamma)\}}{\theta^T_{-M} - \theta + \mathrm{i}\gamma}, \quad (8.4.19)$$

and for intermediate angles $|\mathscr{H}^{(-)}_M(\theta)|$ is much smaller than $|\mathscr{H}^{(+)}_M(\theta)|$. The maximum angular distribution is at $\theta = \theta^T_{-M}$, where the value of the cross-section is

$$\sigma^{(T)}_{LM}(\max) = A\frac{\mu_i\mu_f}{2\pi^2\hbar^4\zeta^2k_i^3k_f\kappa^2}\{d^L_M(\tfrac{1}{2}\pi)\}^2\frac{\Lambda^T_{-M}}{\sin\theta^T_{-M}}$$
$$\times\{I^{(\kappa)}_{LM}(\theta^T_{R,-M},\xi)\}^2\exp(2\gamma\delta_T)\frac{|F(i\Delta\gamma)|^2}{\gamma^2}. \tag{8.4.20}$$

If for example $F(z) = \pi z/\sinh(\pi z)$, we obtain

$$\frac{|F(i\Delta\gamma)|^2}{\gamma^2} = \left\{\frac{\pi\Delta}{\sin(\pi\Delta\gamma)}\right\}^2. \tag{8.4.21}$$

With the same parametric form we have

$$\frac{\mathcal{H}^{(-)}_M(\theta)}{\mathcal{H}^{(+)}_M(\theta)} = \frac{\sinh\{\pi\Delta(\theta^T_{-M}-\theta+i\gamma)\}}{\sinh\{\pi\Delta(\theta^T_M+\theta+i\gamma)\}}$$
$$\approx \exp\{-\pi\Delta(\theta^T_M-\theta^T_{-M})\}\exp(-2\pi\Delta\theta). \tag{8.4.22}$$

This shows that the oscillations in the transfer cross-section are most pronounced at smaller angles and become strongly damped with increasing θ. It is also seen that at a given angle the damping is controlled by the window parameter Δ. Thus if Δ is relatively small, as in the case of 'surface transparency', the oscillations are enhanced. This effect was first discovered by the Brookhaven group (Kahana and Baltz 1977 [8.24]).

Since the radial integrals for the asymptotic form of the transfer form factor (8.2.5) are real functions, we can write the cross-section (8.4.13) more explicitly as

$$\sigma^{(T)}_{LM}(\theta) = \bar{\sigma}^{(T)}_{LM}(\theta)[1+\{\tilde{t}_{LM}(\theta)\}^2+(-)^M 2\tilde{t}_{LM}(\theta)\sin\{2\bar{\Lambda}_M\theta+(\Delta_M-\Delta_{-M})\bar{\theta}\}], \tag{8.4.23}$$

with

$$\tilde{t}_{LM}(\theta) \approx \left(\frac{\Lambda^T_M}{\Lambda^T_{-M}}\right)^{1/2}\frac{I^{(\kappa)}_{L-M}(\theta^T_{R,M},\xi)}{I^{(\kappa)}_{LM}(\theta^T_{R,-M},\xi)}$$
$$\times\exp\{-\pi\Delta(\theta^T_M-\theta^T_{-M})\}\exp(-2\pi\Delta\theta). \tag{8.4.24}$$

This shows that the diffraction oscillations have the angular period $\pi/\bar{\Lambda}_M$, which is M dependent because of

$$\tfrac{1}{2}(\Delta_M+\Delta_{-M}) = \tfrac{1}{2}\Delta\ln\left[\frac{1}{4}\left\{\exp\left(\frac{K_0}{\Delta_i}\right)+\exp\left(-\frac{K_0}{\Delta_f}\right)\right.\right.$$
$$\left.\left.+2\exp\left(\frac{\Delta_f-\Delta_i}{2\Delta_i\Delta_f}K_0\right)\cosh\left(\frac{M}{\Delta}\right)\right\}\right]$$
$$\to\frac{1}{2}\Delta\ln\left[\frac{1}{2}\left\{\cosh\left(\frac{K_0}{\Delta}\right)+\cosh\left(\frac{M}{\Delta}\right)\right\}\right], \tag{8.4.25}$$

with the latter expression for $\Delta_i = \Delta_f$. The M dependence of the relative phase in (8.4.23) is given by

$$\Delta_M - \Delta_{-M} = \Delta \ln\left[\frac{\exp\{(K_0 - M)/2\Delta_i\} + \exp\{-(K_0 - M)/2\Delta_f\}}{\exp\{(K_0 + M)/2\Delta_i\} + \exp\{-(K_0 + M)/2\Delta_f\}}\right]$$
$$\to \Delta \ln\left[\frac{\cosh\{(M - K_0)/2\Delta\}}{\cosh\{(M + K_0)/2\Delta\}}\right]$$
$$\approx \tfrac{1}{2}(|M - K_0| - |M + K_0|). \tag{8.4.26}$$

This can modify the transfer analogue of the Blair phase rule which would otherwise hold because of the factor $(-)^M$ in (8.4.23).

8.4.2 *Magnetic substates and Q-value dependence*

The dependence of the transfer reaction on the kinematic conditions can best be seen from the transfer partial-wave amplitudes (8.2.29)

$$c_{LK}(\bar{\lambda}) = \eta_K(\bar{\lambda}) I^{(\kappa)}_{L-K}(\alpha, \xi) = \eta_K(\bar{\lambda}) \tilde{I}^{(\kappa)}_{L-K}(\alpha, \xi) \exp(-\gamma\bar{\lambda}) \tag{8.4.27}$$

which describe the localization in $\bar{\lambda}$-space. As we have remarked earlier, these functions are essentially asymmetric about their maxima at $\bar{\lambda} = \bar{\Lambda}^T_K$: the fall-off rate at large $\bar{\lambda}$-values is determined by γ and thus by the binding parameter κ, while the rate of decrease at low $\bar{\lambda}$-values is determined by the absorption and measured by Δ^{-1}.

The maximum of $c_{LK}(\bar{\lambda})$ is given (approximately) by

$$c_{LK}(\max) \approx c_{LK}(\bar{\Lambda}^T_K) = \eta_K(\bar{\Lambda}^T_K) I^{(\kappa)}_{L-K}(\theta^T_{R,K}, \xi)$$
$$= \hat{\eta}\left(\frac{\delta_T}{\Delta}\right) \exp(-\gamma\delta_T) \tilde{I}^{(\kappa)}_{L-K}(\theta^T_{R,K}, \xi) \exp(-\gamma\Lambda_K). \tag{8.4.28}$$

Using expression (8.2.19) for Λ_K yields

$$c_{LK}(\max) = \frac{\hat{\eta}(\delta_T/\Delta)\exp\{-\gamma(\bar{\Lambda} + \delta_T)\}\tilde{I}^{(\kappa)}_{L-K}(\theta^T_{R,K}, \xi)}{(\frac{1}{2}[\exp\{(K - K_0)/2\Delta_f\} + \exp\{-(K - K_0)/2\Delta_i\}])^{\gamma\Delta}}$$
$$\approx 2^{\gamma\Delta}\hat{\eta}\left(\frac{\delta_T}{\Delta}\right)\exp\{-\gamma(\bar{\Lambda} + \delta_T)\}\tilde{I}^{(\kappa)}_{L-K}(\theta^T_{R,K}, \xi)\exp\left(-\frac{|K - K_0|}{g^*}\right) \tag{8.4.29}$$

where

$$g^* = \frac{\Delta_i + \Delta_f}{\gamma\Delta_{i,f}} \tag{8.4.30}$$

and the subscript i or f depends on whether $K - K_0$ is positive or negative.

It is instructive to compare the $\bar{\lambda}$- and K-dependence of $c_{LK}(\bar{\lambda})$ with the phenomenological ansatz used by Kahana *et al.* (1974 [8.23]), who assume the Gaussian form

$$c_{LK}^{(\text{phen})}(\bar{\lambda}) = I_L \exp\left\{-\left(\frac{\lambda_i-\Lambda_i}{\Gamma_i}\right)^2 - \left(\frac{\lambda_f-\Lambda_f}{\Gamma_f}\right)^2 - \left(\frac{K-K_0}{g_0}\right)^2\right\}$$
$$\approx I_L \exp\left\{-2\left(\frac{\bar{\lambda}-\bar{\Lambda}}{\Gamma}\right)^2 - \left(\frac{K-K_0}{g}\right)^2\right\}, \qquad (8.4.31)$$

where the second expression is obtained for $\Gamma_i \approx \Gamma_f \approx \Gamma$ and where

$$g = \frac{2^{1/2} g_0 \Gamma}{(g_0^2 + 2\Gamma^2)^{1/2}}. \qquad (8.4.32)$$

Although in our $c_{LK}(\bar{\lambda})$ the asymptotic dependence on $\bar{\lambda}$ and K is exponential rather than Gaussian, there is a correspondence between the parameter g and our $g^* \approx 2/\gamma$, while the parameter Γ corresponds to $\sqrt{2}/\gamma$ for large $\bar{\lambda}$ and to $\sqrt{2}\,\Delta$ for small $\bar{\lambda}$, or rather to the harmonic mean $\Gamma^* = \sqrt{2}\,\Delta/(1+\gamma\Delta)$.

Since the DWBA-based expression (8.4.27) has features which are qualitatively similar to those of the phenomenological form (8.4.31), we have in the present treatment removed the shortcomings of our earlier transfer formalism (Frahn and Sharaf 1969 [8.16]) which were rightly criticized by Kahana and coworkers (Kahana *et al.* 1974 [8.23]; Kahana and Baltz 1977 [8.24]) for ignoring the significant K-dependence of the partial-wave amplitude. The salient point where this is brought out here is the determination of the mean S-matrix (8.2.16) by the method described in Appendix 1. Most of the properties of transfer reactions discussed by Kahana and Baltz (1977 [8.24]) by means of (8.4.31) are described by the present formulation.

To display the M-dependence of the transfer cross-section (in the coordinate frame with the quantization axis perpendicular to the reaction plane) we consider the maximum value (8.4.20) written as

$$\sigma_{LM}^{(T)}(\max) = \hat{A}\{d_{0M}^L(\tfrac{1}{2}\pi)\}^2 \frac{\Lambda_{-M}^T}{\sin\theta_{-M}^T}\{I_{LM}^{(\kappa)}(\theta_{R,-M}^T, \xi)\}^2 \exp(2\gamma\delta_T)$$
$$= \hat{A}\{d_{0M}^L(\tfrac{1}{2}\pi)\}^2 \frac{\Lambda_{-M}^T}{\sin\theta_{-M}^T}\{\tilde{I}_{LM}^{(\kappa)}(\theta_{R,-M}^T, \xi)\}^2 \exp(-2\gamma\Lambda_{-M}).$$
$$(8.4.33)$$

where

$$\hat{A} = A\frac{\mu_i\mu_f}{2\pi^2\hbar^4\zeta^2k_i^3k_f\kappa^2}\frac{|F(i\Delta\gamma)|^2}{\gamma^2}. \qquad (8.4.34)$$

The M- and Q-dependence of the functions $\tilde{I}_{LM}^{(\kappa)}$ is described by the expressions given in Appendix 2. As discussed there, it is contained mainly in a factor $\exp(-b_{-M}^2/2\rho\varepsilon_{-M})$, where

$$b_{-M}=\xi_0'\varepsilon_{-M}+\xi+M\left(\frac{\varepsilon_{-M}-1}{\varepsilon_{-M}+1}\right)^{1/2}$$
$$\varepsilon_{-M}=\{(\Lambda_{-M}^{T}/n)^2+1\}^{1/2}. \tag{8.4.35}$$

However, the most significant factor in (8.4.33) is

$$\begin{aligned}\exp(-2\gamma\Lambda_{-M})&=\exp(-2\gamma\bar{\Lambda})\exp(-2\gamma\Delta_{-M})\\ &=\frac{\exp(-2\gamma\bar{\Lambda})}{(\tfrac{1}{2}[\exp\{(M-M_0)/2\Delta_i\}+\exp\{-(M-M_0)/2\Delta_f\}])^{2\gamma\Delta}}\\ &\approx 4^{\gamma\Lambda}\exp(-2\gamma\Delta)\exp\left(-2\frac{|M-M_0|}{g^*}\right),\end{aligned} \tag{8.4.36}$$

where $M_0=-K_0=\Lambda_f-\Lambda_i$ and g^* is defined by (8.4.30) with index i or f depending on whether $M>M_0$ or $M<M_0$. For $\Delta_i=\Delta_f$ the M-dependence has the simple form $\exp(-\gamma\,|M-M_0|)$, showing that the cross-section is largest for $M=M_0$.

Finally we consider the M-dependence of the amplitude (8.4.24) of the diffraction oscillations, which can be written as

$$\tilde{t}_{LM}(\theta)=\left(\frac{\Lambda_M^T}{\Lambda_{-M}^T}\right)^{1/2}\frac{\tilde{I}_{L-M}^{(\kappa)}(\theta_{R,M}^T,\xi)}{\tilde{I}_{LM}^{(\kappa)}(\theta_{R,-M}^T,\xi)}$$
$$\times\exp\{-\gamma(\bar{\Lambda}_M^T-\bar{\Lambda}_{-M}^T)\}\exp\{-\pi\Delta(\theta_M^T-\theta_{-M}^T)\}\exp\{-2\pi\Delta\theta\}. \tag{8.4.37}$$

Since $\Delta_{\pm M}\mp\frac{1}{2}M+\delta_T\ll\bar{\Lambda}$ and $\bar{\Lambda}_M^T-\bar{\Lambda}_{-M}^T=\Delta_M-\Delta_{-M}$, and further

$$\theta_M^T-\theta_{-M}^T=\Theta(\bar{\Lambda}_M^T)-\Theta(\bar{\Lambda}_{-M}^T)\approx(\Delta_M-\Delta_{-M})\Theta'(\bar{\Lambda}), \tag{8.4.38}$$

this becomes

$$\tilde{t}_{LM}(\theta)\approx\frac{\tilde{I}_{L-M}^{(\kappa)}(\theta_{R,M}^T,\xi)}{\tilde{I}_{LM}^{(\kappa)}(\theta_{R,-M}^T,\xi)}\exp[-\{\gamma+\pi\Delta\Theta'(\bar{\Lambda})\}(\Delta_M-\Delta_{-M})]\exp(-2\pi\Delta\theta), \tag{8.4.39}$$

where $\Delta_M-\Delta_{-M}$ is given by (8.4.26). Apart from this difference, because $\Theta'(\bar{\Lambda})<0$ the sign of the exponential depends on the relative magnitude of γ and $\pi\Delta\,|\Theta'(\bar{\Lambda})|$. An estimate of the latter quantity is given by the Rutherford value $\pi(\Delta/\Lambda)\sin\bar{\theta}_R$ which under strong-absorption conditions is small compared with γ. Thus the amplitudes are largest for good l-matching (small K_0) in which case $\Delta_M-\Delta_{-M}$ is almost independent of M, and the main M-dependence of $t_{LM}(\theta)$ is given by the ratio of the functions $I_{LM}^{(\kappa)}$. Since from Appendix 2 we know that these are largest for

$M=-L$ and smallest for $M=L$, the amplitude is a maximum for $M=L$. Further, the magnitude of $t_{LL}(\theta)$ decreases with increasing L, so that the strongest diffraction oscillations occur for $L=0$ transfer and are weaker for higher multipolarities.

8.5 Conclusion and outlook

The expressions derived in this Chapter give, in closed analytic form, a rather detailed description of the significant wave-mechanical aspects of heavy-ion transfer reactions as embodied implicitly in (first-order no-recoil) DWBA calculations. By accounting for the important kinematical and dynamical effects of magnetic substate, angular momentum matching, and Q-value dependences, a considerable improvement is achieved over earlier closed-form evaluations of the DWBA amplitude which glossed over many subtle features by taking averages over initial and final state energies and angular momentum projections that were too rough. The main improvement is the introduction of the 'mean S-matrix', which leads to an explicit representation of the partial-wave transfer amplitude that is more realistic than the *ad hoc* forms assumed in phenomenological treatments.

Of course, the formalism presented so far still has many shortcomings because of the simplifying assumptions made at the outset: (i) the no-recoil approximation, (ii) omission of higher-order and multistep processes, and (iii) the neglect of spin and polarization effects. In subsequent work we endeavour to remove or remedy these limitations by suitable extensions of our method. One of several ways to account for recoil effects is a Taylor expansion of the final-channel distorted wave function (Nagurajan 1972 [8.35]; Baltz and Kahana 1974 [8.36]), which results in the no-recoil transfer amplitude being amended by a linear combination of amplitudes pertaining to different values L_R of the transferred angular momentum and containing radial integrals with modified transfer form factors. The additional amplitudes can then be evaluated in closed form using similar approximations to those adopted for the no-recoil term. Further, the contributions of higher-order processes, such as inelastic scattering of the initial and final nuclei before and after the transfer takes place, can be taken into account by replacing the initial and final S-matrices with the modified expressions derived recently (Frahn and Hussein, 1980 [8.37]) from a coupled-channels extension of the closed-formalism for elastic and inelastic heavy-ion scattering. Earlier, Udagawa and Tamura (1975 [8.38]) gave a qualitative description by using a phenomenological form of the partial-wave transfer amplitude. Similar extensions are possible for multistep, successive, or sequential transfers and will be treated in subsequent papers. Lastly, spin and polarization effects can be described by including spin–orbit interaction in the initial and final S-matrices and taking the spin dependence of the bound-state wave function into account.

8.6 References

[8.1] Frahn, W. E. and Venter, R. H. (1964). *Nucl. Phys.* **59,** 651.
[8.2] Strutinsky, V. M. (1964). *Sov. Phys.—JETP* **19,** 1401.
[8.3] Strutinsky, V. M. (1973). *Phys. Lett. B* **44,** 245.
[8.4] Dodd, L. R. and Greider, K. R. (1965). *Phys. Rev. Lett.* **14,** 959.
[8.5] Dodd, L. R. and Greider, K. R. (1969). *Phys. Rev.* **180,** 1187.
[8.6] Christensen, P. R., Hansen, O., Larsen, J. S., Sinclair, D., and Videbaek, F. (1973). *Phys. Lett. B* **45,** 107.
[8.7] Bond, P. D., Garrett, J. D., Hansen, O., Kahana, S., LeVine, M. J., and Schwarzschild, A. Z., (1973). *Phys. Lett.* **47,** 231.
[8.8] Lemaire, M. C., Mermaz, M. C., Sztark, H., and Cunsolo, A. (1974). *Phys. Rev. C* **10,** 1103.
[8.9] LeVine, M. J., Baltz, A. J., Bond, P. D., Garrett, J. D., Kahana, S., and Thorn, C. E. (1974). *Phys Rev. C* **10,** 1602.
[8.10] Henning, W., Kovar, D. G., Zeidman, B., and Erskine, J. R. (1974). *Phys. Rev. Lett.* **32,** 1015.
[8.11] Dar, A. (1965). *Phys. Rev. B* **139,** 1193.
[8.12] Dar, A. (1966). *Nucl. Phys.* **82,** 354.
[8.13] Dar, A. and Kozlowsky, B. (1965). *Phys. Rev. Lett.* **15,** 1036.
[8.14] Varma, S. (1968). *Nucl. Phys. A* **106,** 233.
[8.15] Frahn, W. E. and Venter, R. H. (1963). *Ann. Phys. (N.Y.)* **24,** 243.
[8.16] Frahn, W. E. and Sharaf, M. A. (1969). *Nucl. Phys. A* **133,** 593.
[8.17] Broglia, R. A. and Winther, Aa. (1972). *Phys. Rep.* **4C,** 153.
[8.18] Landowne, S., Dasso, C. H., Nilsson, B. S., Broglia, R. A., and Winther, Aa. (1976). *Nucl. Phys. A* **259,** 99.
[8.19] Knoll, J. and Schaeffer, R. (1977). *Phys. Rep.* **31C,** 159.
[8.20] Hasan, H. and Brink, D. M. (1978). *J. Phys. G* **4,** 1573.
[8.21] Chasman, C., Kahana, S., and Schneider, M. (1973). *Phys. Rev. Lett.* **31,** 1074.
[8.22] Schneider, M., Chasman, C., Kahana, S., Baltz, A. J., and Auerbach, E. H. (1973). *Phys. Rev. Lett.* **31,** 320.
[8.23] Kahana, S., Bond, P. D., and Chasman, C. (1974). *Phys. Lett. B* **50,** 199.
[8.24] Kahana, S. and Baltz, A. J. (1977). *Adv. nucl. Phys.* **9,** 1.
[8.25] Frahn, W. E. (1975). In *Heavy-ion, high-spin states and nuclear structure,* Vol. 1, p. 157. International Atomic Energy Agency, Vienna.
[8.26] Frahn, W. E. (1975). In *Classical and quantum mechanical aspects of heavy ion collisions. Lecture notes in physics,* Vol. 33, p. 102. Springer, Berlin.
[8.27] Frahn, W. E. (1976). *Nucl. Phys. A* **272,** 413.
[8.28] Austern, N. (1970). *Direct nuclear reaction theories.* John Wiley, New York.

[8.29] Sopkovich, M. J. (1962). *Nuovo Cimento* **26,** 186.
[8.30] Gottfried, K. and Jackson, J. D. (1964). *Nuovo Cimento* **34,** 735.
[8.31] Trautmann, D. and Alder, K. (1970). *Helv. Phys. Acta* **43,** 363.
[8.32] Alder, K. and Trautmann, D. (1971). *Ann. Phys.* (*N.Y.*) **66,** 884.
[8.33] Alder, K., Morf, R., Pauli, M., and Trautmann, D. (1972). *Nucl. Phys. A* **191**, 399.
[8.34] Frahn, W. E. (1980). *Nucl. Phys. A* **337,** 324.
[8.35] Nagarajan, M. A. (1972). *Nucl. Phys. A* **196,** 34.
[8.36] Baltz, A. J. and Kahana, S. (1974). *Phys. Rev. C* **9,** 2243.
[8.37] Frahn, W. E. and Hussein, M. S. (1980). *Nucl. Phys. A* **346,** 237.
[8.38] Udagawa, T. and Tamura, T. (1975). *Phys. Lett. B* **57,** 135.

9

COUPLED-CHANNEL EXTENSIONS

9.1 Introduction

It is well known that in heavy-ion collisions channel coupling often has important effects on elastic scattering, inelastic scattering, and transfer reactions. These effects manifest themselves in deviations from the 'normal' shapes of angular distributions, in anomalous large-angle scattering, and in unusual structures of excitation functions. Since customary optical model and DWBA calculations and simple S-matrix descriptions are found to be inadequate in these cases, the use of much more elaborate coupled-channel computations appears to be inevitable. Because these can be prohibitively expensive and time consuming, it is highly desirable to devise methods by which channel-coupling effects can be represented in a form, albeit very approximate, that can be used in straightforward extensions of the conventional calculations.

One such method is to derive effective potentials which can be added to the normal optical potential to simulate the feedback of strongly coupled channels on the elastic scattering. Since the effective interaction is non-local, it is usually approximated by an 'equivalent' local potential which then, in general, depends on the orbital angular momentum l. Such potentials for the case of coupling to low-lying collective (2^+) states via Coulomb excitation have been derived by several authors (Love *et al.* 1977 [9.1]; Love *et al.* 1977 [9.2]; Baltz *et al.* 1978 [9.3]; Donangelo *et al.* 1979 [9.4]). As a by-product of the method described in this Chapter, Frahn and Hussein (1980 [9.5]) have derived more general effective potentials which represent coupling to inelastic channels by both Coulomb and nuclear excitation and coupling to transfer channels.

In the present chapter we develop a procedure which enables us, starting from a set of coupled-channel equations and using suitable approximations, to calculate in explicit form the modifications of the elastic S-matrix that arise from coupling to certain types of strongly excited channels. These modifications are determined solely by the characteristics of the transitions to the intermediate states (i.e. the spectroscopic factors and the form factors) and by the uncoupled elastic S-matrix. Their analytic forms are such that the corresponding contributions to the scattering amplitude can be evaluated in closed form by the same methods as developed in previous chapters (Frahn 1975 [9.6]; Frahn and Gross (1976) [9.7]; Frahn 1976 [9.8]; Frahn and Rehm 1978 [9.9]; Frahn 1980 [9.10]). In this way we achieve a natural and straightforward extension of the closed formalism (CF) to account for channel-coupling effects.

We consider it a significant feature of our approach that the modifications of the uncoupled S-matrix, and hence of the scattering amplitude, are calculated directly without the need to determine explicitly the effective interactions which appear only in an intermediary role in our derivation. Thereby we avoid, in contrast with earlier work (Love *et al.* 1977 [9.1]; *Love et al.* 1977 [9.2]; Baltz *et al.* 1978 [9.3]; Donangelo *et al.* 1979 [9.4]; Frahn and Hussein 1980 [9.5]), the two-step procedure of first calculating an effective potential and then solving the Schrödinger equation to calculate the scattering amplitude. Moreover, whereas the latter can only be done numerically, the analytic form of our results displays explicitly the physical nature and dynamical origin of the various contributions to the scattering cross-section.

For simplicity we consider only individual channels and treat the coupling in lowest order, but we show for some cases how the method can be extended to higher-order multistep processes by iteration procedures. As specific examples we consider the most important channels: inelastic scattering (by Coulomb and nuclear excitation) and particle transfer (including one-step and two-step elastic transfer processes). Most of the material of this chapter is based on the work of Frahn and Hussein (1980 [9.11]).

9.2 Generic formulation

We start from the coupled-channel equations for the radial wave functions $\chi_{(lI)J}(k, r)$ for orbital angular momentum l, spin I, and total angular momentum J in channels denoted by n, m:

$$\left\{\frac{\mathrm{d}^2}{\mathrm{d}r^2}+k_n^2-\frac{l_n(l_n+1)}{r^2}-\frac{2\mu_n}{\hbar^2}\mathring{U}(r)\right\}\chi_{(l_nI_n)J}(k_n, r)\exp\{\mathrm{i}\sigma_{l_n}(k_n)\}$$
$$=\sum_{l_mI_m}\frac{2\mu_n}{\hbar^2}V^J_{l_nI_n,l_mI_m}(r)\chi_{(l_mI_m)J}(k_m, r)\exp\{\mathrm{i}\sigma_{l_m}(k_m)\}, \quad (9.2.1)$$

where μ is the reduced mass and the $\sigma_l(k)$ are Rutherford phase shifts. For simplicity we consider spin-zero nuclei in the entrance channel which we denote by n, i.e. $I_n = 0$. This fixes the value of J to be l_n. The interaction in the elastic channel in the absence of coupling to the channels m is assumed to be described by a local optical potential $U_{\mathrm{N}}(r)$

and the Coulomb potential $V_C(r)$:

$$\mathring{U}(r) = V_c(r) + U_N(r) \qquad U_N(r) = V(r) + iW(r). \tag{9.2.2}$$

We write the coupling interaction as

$$V^{l_n}_{l_n0,l_mL}(r) \equiv V^L_{l_nl_m}(r) = a^{(m)}_{l_nl_m} F^{(m)}_L(r), \tag{9.2.3}$$

where $F^{(m)}_L(r)$ is the form factor for the transition $n \to m$ with multipolarity L, and all spectroscopic and geometric factors are lumped together in $a^{(m)}_{l_nl_m}$.

Since the nonelastic channels $m \neq n$ have outgoing waves only, the wave functions $\chi_{l_m}(k_m, r)$ (in simplified notation) satisfy the integral equations

$$\chi_{l_m}(k_m, r)\exp(i\sigma_{l_m}) = \frac{2\mu_n}{\hbar^2} \int_0^\infty dr' \mathring{G}^{(+)}_{l_m}(r, r') \times \sum_{l_s} V^L_{l_ml_s}(r')\chi_{l_s}(k_s, r')\exp(i\sigma_{l_s}), \tag{9.2.4}$$

where $\mathring{G}^{(+)}_{l_s}(r, r')$ is the outgoing-wave Green function for the left-hand side of eqn (9.2.1).

Our first approximation is to neglect all couplings except to the elastic channel:

$$\chi_{l_m}(k_m, r)\exp(i\sigma_{l_m}) \approx \frac{2\mu_n}{\hbar^2} \int_0^\infty dr' \mathring{G}^{(+)}_{l_m}(r, r') V^L_{l_ml_n}(r')\chi_{l_n}(k_n, r')\exp(i\sigma_{l_n}). \tag{9.2.5}$$

Insertion of (9.2.5) in (9.2.1) results in a Schrödinger equation for $\chi_{l_n}(k_n, r)$ with an effective interaction

$$U_{l_n}(r) = \mathring{U}(r) + \tilde{U}_{l_n}(r), \tag{9.2.6}$$

where

$$\tilde{U}_{l_n}(r)\chi_{l_n}(k_n, r) = \frac{2\mu_n}{\hbar^2} \sum_{l_m} V^L_{l_nl_m}(r) \int_0^\infty dr' \mathring{G}^{(+)}_{l_m}(r, r') V^L_{l_ml_n}(r')\chi_{l_n}(k_n, r'). \tag{9.2.7}$$

We bypass an explicit determination of the (non-local) interaction operator $\tilde{U}_{l_n}(r)$ by calculating directly the partial-wave S-matrix elements. From the radial Gell-Mann–Goldberger relation we obtain for the nuclear part of the total S-matrix $S_l = S^{(N)}_l \exp(i2\sigma_l)$,

$$S^{(N)}_{l_n}(k_n) = \mathring{S}^{(N)}_{l_n}(k_n) + \tilde{S}^{(N)}_{l_n}(k_n), \tag{9.2.8}$$

where $\mathring{S}_l$ is the uncoupled elastic S-matrix and

$$\tilde{S}^{(n)}_{l_n}(k_n) = -i\frac{4\mu_n}{\hbar^2 k_n} \int_0^\infty dr \mathring{\chi}_{l_n}(k_n, r)\tilde{U}_{l_n}(r)\chi_{l_n}(k_n, r) \tag{9.2.9}$$

gives the channel-coupling contribution. Insertion of (9.2.7) yields

$$\tilde{S}^{(\mathrm{N})}_{l_n}(k_n) = -\mathrm{i}\frac{8\mu_n^2}{\hbar^4 k_n}\sum_{l_m}\int_0^\infty \mathrm{d}r\mathring{\chi}_{l_n}(k_n, r)V^L_{l_n l_m}(r) \times \int_0^\infty \mathrm{d}r'\mathring{G}^{(+)}_{l_m}(r, r')V^L_{l_m l_n}(r')\chi_{l_n}(k_n, r'). \tag{9.2.10}$$

Our next approximation is to replace χ_{l_n} on the right-hand side of (9.2.10) by the uncoupled wave function $\mathring{\chi}_{l_n}$. This amounts to treating the coupling to lowest order; higher-order contributions can be obtained by iteration. Our third step is to approximate the Green function in channel m by its on-energy-shell part

$$\mathring{G}^{(+)}_{l_m}(r, r') \approx -\frac{\mathrm{i}}{k_m}\mathring{\chi}_{l_m}(k_m, r)\mathring{\chi}_{l_m}(k_m, r')\{\mathring{S}^{(\mathrm{N})}_{l_m}(k_m)\}^{-1}. \tag{9.2.11}$$

Using (9.2.3) we can then write eqn (9.2.10) as

$$\tilde{S}^{(\mathrm{N})}_{l_n}(k_n) = -\frac{\mu_n^2 k_m k_n}{2\pi^2\hbar^4} \times \sum_{l_m} a^{(n)}_{l_m l_n} a^{(m)}_{l_n l_m} R^{(n)L}_{l_n l_m}(k_n, k_m) R^{(m)L}_{l_m l_n}(k_m, k_n)\{\mathring{S}^{(\mathrm{N})}_{l_m}(k_m)\}^{-1}, \tag{9.2.12}$$

where

$$R^{(m)L}_{l_m l_n}(k_m, k_n) = \frac{4\pi}{k_m k_n}\int_0^\infty \mathrm{d}r\mathring{\chi}_{l_m}(k_m, r)F^{(m)}_L(r)\mathring{\chi}_{l_n}(k_n, r) \tag{9.2.13}$$

are the DWBA radial integrals for the transition $n \to m$ with multipolarity L.

The total elastic S-matrix can now be written in the form

$$S_{l_n}(k_n) = \{1 - t^{(n,m)}_{l_n}(k_m, k_n)\}\mathring{S}_{l_n}(k_n) \tag{9.2.14}$$

with

$$t^{(n,m)}_{l_n}(k_m, k_n) = \frac{\mu_n^2 k_m k_n}{2\pi^2\hbar^4}\sum_{l_m} a^{(n)}_{l_m l_n} a^{(m)}_{l_n l_m} T^{(n,m)}_{l_m l_n}(k_m, k_n), \tag{9.2.15}$$

where we have defined

$$T^{(n,m)}_{l_m l_n}(k_m, k_n) = \frac{R^{(n)L}_{l_n l_m}(k_n, k_m) R^{(m)L}_{l_m l_n}(k_m, k_n)}{\mathring{S}^{(\mathrm{N})}_{l_m}(k_m)\mathring{S}^{(\mathrm{N})}_{l_n}(k_n)}. \tag{9.2.16}$$

In fact, the use of the on-energy-shell approximation of the Green function (eqn (9.2.11)) allows us to perform a summation over terms of all orders in $t^{(n,m)}_{l_n}(k_m, k_n)$ in the form of a geometrical (Born-type) series, with the result

$$S_{l_n}(k_n) = \frac{1 - \frac{1}{2} t_{l_n}^{(n,m)}(k_m, k_n)}{1 + \frac{1}{2} t_{l_n}^{(n,m)}(k_m, k_n)} \mathring{S}_{l_n}(k_n). \tag{9.2.17}$$

Note that because of the coupling $\mathbf{l}_m + \mathbf{L} = \mathbf{l}_n$ only a few (at most $2L+1$) terms contribute to the sum in (9.2.15).

We can regard eqn (9.2.14), or the iterated form (9.2.17), with the expressions (9.2.15) and (9.2.16) as our main result: it gives the coupling corrections to the S-matrix in the form of well-defined expressions determined by the DWBA radial integrals and the unperturbed S-matrices.

However, in order to proceed towards analytic expressions and lay the ground for closed-form evaluation of the scattering amplitude, we employ for the examples treated in the following sections further approximations that have been used in the CF discussed earlier.

9.3 Dynamic polarization

The effect of the coupling to (low-lying collective) inelastic channels on elastic scattering is termed 'dynamic polarization'. For a given channel of spin $I_m = L$ this is, to lowest order, a two-step process in which the level in question is excited and de-excited by inelastic transitions of multipolarity L. This effect has been studied for Coulomb excitation; the effective potential method was used by Love *et al.* (1977 [9.1]), Love *et al.* (1977 [9.2]), Baltz *et al.* (1978 [9.3]), and Donangelo (1979 [9.4]), and the extended CF was used by Frahn and Hill (1978 [9.12]). In our present treatment we also include the effects of nuclear excitation.

In the 'extended optical model' the coupling interaction (9.2.3) is

$$V_{ll'}^{L}(r) = a_{ll'}^{L} F_L(r), \tag{9.3.1}$$

where

$$a_{ll'}^{L} = \frac{5}{2\pi^{1/2}} \left\{ \frac{(2l+1)(2l'+1)}{2L+1} \right\}^{1/2} (-)^l \begin{Bmatrix} l & l' & L \\ L & 0 & l \end{Bmatrix} \begin{pmatrix} l & l' & L \\ 0 & 0 & 0 \end{pmatrix} \tag{9.3.2}$$

and the form factor is given by

$$F_L(r) = \delta_L^{(C)} \frac{3}{2L+1} Z_A Z_B c^2 \frac{R_c^{L-1}}{r^{L+1}} - \delta_L^{(N)} \frac{dU_N(r)}{dr} \tag{9.3.3}$$

for $r \geq R_c$, where R_c is the charge radius. (Here and in all following expressions we write $l_n = l$ and $l_m = l'$.) The Coulomb and nuclear deformation lengths are denoted by $\delta_L^{(C)}$ and $\delta_L^{(N)}$ respectively, and Z_A and Z_B are the atomic numbers of the scattering nuclei A and B respectively.

Now we use the same approximations for the radial integrals (9.2.13) as were used by Frahn (1976 [9.8] and Frahn and Rehn (1978 [9.9]) in the CF for inelastic scattering:

$$R^{L}_{l'l}(k',k)=\frac{4\pi}{k'k}\int_0^\infty \mathrm{d}r\mathring{\chi}_{l'}(k',r)F_L(r)\mathring{\chi}_l(k,r)$$
$$\approx\frac{\pi\hbar^2}{\mu}\left[\delta^{(C)}_L\hat{C}_{L-K}(\bar{\lambda})\{\mathring{S}^{(N)}_{l'}(k')\mathring{S}^{(N)}_{l}(k)\}^{1/2}\right.$$
$$\left.-\mathrm{i}\delta^{(N)}_L\left\{\frac{\mathrm{d}\mathring{S}^{(N)}_{l'}(k')}{\mathrm{d}l'}\frac{\mathrm{d}\mathring{S}^{(N)}_{l}(k)}{\mathrm{d}l}\right\}^{1/2}\right], \tag{9.3.4}$$

where

$$\hat{C}_{L-K}(\bar{\lambda})=\frac{3}{2L+1}\left(\frac{\bar{k}R_c}{\bar{n}}\right)^{L-1}I_{L-K}(\vartheta,\xi). \tag{9.3.5}$$

We have written $k_n=k$, $k_m=k'$, and $\mu_n=\mu$, and have defined

$$K=l-l',\qquad \bar{\lambda}=\bar{l}+\tfrac{1}{2},\qquad \bar{l}=\tfrac{1}{2}(l+l'),$$
$$\bar{n}=\tfrac{1}{2}(n+n'),\qquad \xi=n'-n,\qquad \vartheta=2\arctan(\bar{n}/\bar{\lambda}), \tag{9.3.6}$$

where n and n' are the Sommerfeld parameters in the initial and final channels respectively. The functions $I_{L-K}(\vartheta,\xi)$ are defined as by Frahn (1976 [9.8]) and Frahn and Rehm (1978 [9.9]) as standard WKB radial integrals for Coulomb excitation. For details of the approximations leading to eqn (9.3.4) (which involve the 'Sopkovich approximation' for the Coulomb part of the radial integrals, the WKB approximation for the Coulomb excitation integrals, and the Austern–Blair relation with Hahne's modification for the nuclear part of the radial integrals) we refer to Frahn (1976 [9.8]) and Frahn and Rehm (1978 [9.9]) and the references given therein.

It should be noted that expression (9.3.4) is symmetrical in the initial and final channels. Thus we obtain for the quantities defined by eqn (9.2.16)

$$T_{l'l}(k',k)=\frac{\{R^{L}_{l'l}(k',k)\}^2}{\mathring{S}^{(N)}_{l'}(k')\mathring{S}^{(N)}_{l}(k)}$$
$$=\left(\frac{\pi\hbar^2}{\mu}\right)^2\{\delta^{(C)}_L\hat{C}_{L-K}(\bar{\lambda})-\mathrm{i}\delta^{(N)}_L\hat{N}_k(\bar{\lambda})\}^2, \tag{9.3.7}$$

with the definition

$$\hat{N}_K(\bar{\lambda})=\left\{\frac{\mathrm{d}\mathring{S}^{(N)}_{l'}(k')/\mathrm{d}l'}{\mathring{S}^{(N)}_{l'}(k')}\cdot\frac{\mathrm{d}\mathring{S}^{(N)}_{L}(k)/\mathrm{d}l}{\mathring{S}^{(N)}_{l}(k)}\right\}^{1/2}. \tag{9.3.8}$$

Since the coefficients $a^L_{ll'}$ are also symmetrical in l and l', our result for the total elastic S-matrix, modified by dynamic polarization, can be written in the form (9.2.14)

$$S_l(k) = \{1 - t_l(k', k)\}\mathring{S}_l(k) \tag{9.3.9}$$

with

$$\begin{aligned} t_l(k', k) = \tfrac{1}{2}kk' \sum_{K=-L}^{L} \{a_{LK}(\bar{\lambda})\}^2 \\ \times[(\delta_L^{(c)})^2\{\hat{C}_{L-K}(\bar{\lambda})\}^2 - (\delta_L^{(N)})^2\{\hat{N}_k(\bar{\lambda})\}^2 \\ - \mathrm{i}2\delta_L^{(C)}\delta_L^{(N)}\hat{C}_{L-K}(\bar{\lambda})\hat{N}_K(\bar{\lambda})], \end{aligned} \tag{9.3.10}$$

where we have written $a^L_{ll'}\equiv a_{LK}(\bar{\lambda})$ and replaced the summation over $l_m = l'$ by a sum over K.

This expression displays the contributions of dynamic polarization due to Coulomb excitation, nuclear excitation, and the *interference* between Coulomb and nuclear excitation.

Equations (9.3.9) and (9.3.10) give only the lowest-order effect of the dynamic polarization (bilinear in the deformation lengths $\delta_L^{(C)}$ and $\delta_L^{(N)}$). As shown in Section 9.2, a Born-type iteration procedure allows us to sum all higher-order contributions—within our approximations of channel coupling—in the form of a geometrical series, with the result

$$S_l^{(\text{geom})}(k) = \frac{1 - \frac{1}{2}t_l(k', k)}{1 + \frac{1}{2}t_l(k', k)}\mathring{S}_l(k), \tag{9.3.9a}$$

where $t_l(k', k)$ is given by eqn (9.3.10). A different way of including higher-order effects (which we cannot derive within our procedure because it corresponds to a WKB-type or eikonal approximation) leads to an exponential form of the dynamic polarization correction:

$$S_l^{(\exp)}(k) = \mathring{S}_l(k)\exp\{-t_l(k', k)\}. \tag{9.3.9b}$$

Obviously, both expressions (9.3.9*a*) and (9.3.9*b*) reduce to (9.3.9) to lowest order in the coupling constants; in fact they agree even to second order in t_l.

For pure Coulomb excitation and in the adiabatic limit $k' = k$, i.e. with $t_l(k', k) \to t_C(\bar{\lambda})$ given by

$$t_C(\bar{\lambda}) = \tfrac{1}{2}(k\delta_L^{(C)})^2 \sum_{K=-L}^{L} \{a_{LK}(\bar{\lambda})\hat{C}_{L-K}(\bar{\lambda})\}^2, \tag{9.3.10a}$$

the exponential modification of $\mathring{S}_l(k)$ in (9.3.9*b*) is equivalent to that derived by Frahn and Hill (1978 [9.12]) from the l-dependent effective potential of Baltz *et al.* (1978 [9.3]) using the so-called 'Coulomb-distorted eikonal approximation'.

9.4 Strong Coulomb excitation

In several cases, such as the scattering of $^{18}O + ^{18}W$ at 90 MeV (Thorn *et al.* 1977 [9.13]) analysed by Love *et al.* (1977 [9.1]), Love *et al.* (1977 [9.2]), Baltz *et al.* (1978 [9.3]), and Frahn and Hill (1978 [9.12]), the dynamic polarization by Coulomb excitation is so strong that the lowest-order approximation, is insufficient. Then we must use the expressions obtained by summation over higher-order contributions, either in the geometrical (Born) series form (9.3.9*a*) or in the exponential (WKB) form (9.3.9*b*). If we confine ourselves to pure Coulomb excitation ($\delta_L^{(N)} = 0$), the total amplitude becomes

$$f(\theta) \equiv f_C(\theta) = \mathring{f}(\theta) + \tilde{f}_C(\theta), \tag{9.4.1}$$

which is evaluated with the S-matrix

$$S_C^{(\text{geom})}(\lambda) = \frac{1 - \frac{1}{2}t_C(\bar{\lambda})}{1 + \frac{1}{2}t_C(\bar{\lambda})}\mathring{S}(\lambda)$$

or

$$S_C^{(\exp)}(\lambda) = \mathring{S}(\lambda)\exp\{-t_C(\bar{\lambda})\}. \tag{9.4.2}$$

Because of the slow variation of $t_C(\bar{\lambda})$ compared with $\mathring{S}_N(\lambda)$, the corresponding amplitudes $\tilde{f}_C(\theta)$ have t_C (of argument $\bar{\lambda}_\theta$ or $\bar{\Lambda}$) replaced by

$$t_C^{(\text{geom})} = \frac{t_C}{1 + \frac{1}{2}t_C}$$

or

$$t_C^{(\exp)} = 1 - \exp(-t_C). \tag{9.4.3}$$

As the simplest case let us consider scattering below the Coulomb barrier. Then the ratio of the total differential cross-section $\sigma_C(\theta)$ to the Rutherford cross-section $\sigma_R(\theta)$ becomes

$$\frac{\sigma_C^{(\text{geom})}(\theta)}{\sigma_R(\theta)} = \left\{\frac{1 - \frac{1}{2}t_C(\bar{\lambda}_\theta)}{1 + \frac{1}{2}t_C(\bar{\lambda}_\theta)}\right\}^2 \tag{9.4.4}$$

or

$$\frac{\sigma_C^{(\exp)}(\theta)}{\sigma_R(\theta)} = \exp\{-2t_C(\bar{\lambda}_\theta)\}. \tag{9.4.5}$$

As a specific example we give the explicit expression for $t_C(\bar{\lambda}_\theta)$, defined by eqn (9.3.10*a*), for quadrupole excitation. With the coefficients $a_{2K}(\bar{\lambda})$ from eqn (9.3.2), calculated in the large-l limit,

$$a_{20}(\bar{\lambda}) \simeq -\frac{1}{4\pi^{1/2}}, \qquad a_{2\pm 2}(\bar{\lambda}) \simeq \left(\frac{3}{2}\right)^{1/2}\frac{1}{4\pi^{1/2}}, \tag{9.4.6}$$

we obtain

$$t_{\mathrm{C}}(\bar{\lambda}_\theta) \simeq \frac{1}{32\pi}(k\delta_2^{(\mathrm{C})})^2$$

$$\times\left(\{\hat{C}_{20}(\bar{\lambda}_\theta)\}^2 + \frac{3}{2}[\{\hat{C}_{22}(\bar{\lambda}_\theta)\}^2 + \{\hat{C}_{2-2}(\bar{\lambda}_\theta)\}^2]\right)$$

$$\approx \frac{\pi}{50}\frac{k^4}{n^2}\frac{B(E2,\uparrow)}{Z_{\mathrm{B}}^2 e^2}$$

$$\times(\{I_{20}(\theta,\xi)\}^2 + \tfrac{3}{2}[\{I_{22}(\theta,\xi)\}^2 + \{I_{2-2}(\theta,\xi)\}^2]),$$

and on using the explicit expressions (Alder *et al.* (1956 [9.14]) for the functions $I_{2K}(\theta, 0)$ in the adiabatic limit $\xi = 0$ this becomes

$$t_{\mathrm{C}}(\bar{\lambda}_\theta) = \frac{2\pi}{75}\frac{k^4}{n^2}\frac{B(E2,\uparrow)}{Z_{\mathrm{B}}^2 e^2}$$

$$\times[(\sin\tfrac{1}{2}\theta)^4 + 3(\tan\tfrac{1}{2}\theta)^4\{1 - \tfrac{1}{2}(\pi-\theta)\tan\tfrac{1}{2}\theta\}^2]. \qquad (9.4.8)$$

Expressions (9.4.4) and (9.4.8) agree with a result obtained recently by Baltz *et al.* (1979 [9.15]). Equation (9.4.5) is identical with the result obtained by Baltz *et al.* (1978 [9.3]) at $\theta = 180°$; otherwise the explicit angular dependence is slightly different. As long as the coupling is not very strong so that $\frac{1}{2}t_{\mathrm{C}}(\bar{\lambda}_\theta)$ is less than unity for all angles we expect eqns (9.4.4) and (9.4.5) to give comparable results. To exhibit this we show in Fig. 9.1 a comparison of the cross-section ratios (9.4.4) and (9.4.5) with

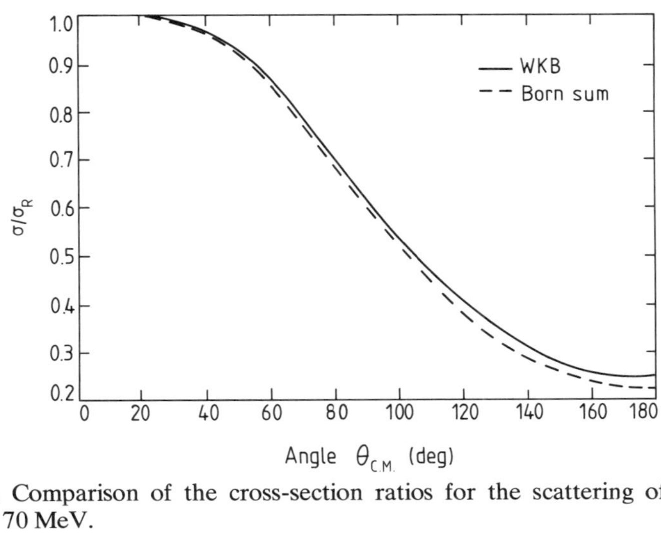

FIG. 9.1. Comparison of the cross-section ratios for the scattering of ^{20}Ne by ^{152}Sm at 70 MeV.

$t_C(\bar{\lambda}_\theta)$ given by eqn (9.4.8) for the scattering of $^{20}Ne + ^{152}Sm$ at $E_{lab} = 70$ MeV (Doll *et al.* (1978 [9.16]). (Reorientation, projectile excitation, etc. are neglected in the expressions, although our formalism could be extended to include these effects as well.)

9.5 Coupling to transfer channels

Now we turn to the coupling to rearrangement channels, and consider processes in which the intermediate channel m differs from the initial channel n by the transfer of a nucleon or a cluster c from one of the scattering nuclei (A, say) to the other (B). Here we encounter the problem of non-orthogonality of the wave functions in the initial and intermediate channels. For the two-step process $A+B \to A'+B' \to A+B$ this results in an additional term in the effective interaction potential for elastic scattering. Although our treatment could be extended to include such terms, we shall for simplicity neglect the non-orthogonality contributions.

Thus we use the results Section 9.2.2 (again writing $l_n = l$, $l_m = l'$, and $\mu_n = \mu$) with the coupling interaction

$$V^L_{ll'}(r) = a^{(\kappa')}_T F^{(\kappa')}_L(r), \tag{9.5.1}$$

where the coefficient $a^{(\kappa')}_T$ includes the product of spectroscopic factors and angular momentum coupling terms, and

$$F^{(\kappa')}_L(r) = -i^L h^{(1)}_L(i\kappa' r) \simeq \frac{\exp(-\kappa' r)}{\kappa' r} \tag{9.5.2}$$

is the form factor for transfer into a bound state of imaginary wave number $\kappa' = (2\mu'_c E'_B)^{1/2}/\hbar$ where μ'_c is the reduced mass and E'_B is the binding energy of the transferred particle c in the nucleus B′.

Now we employ the approximations used in the closed-form description of heavy-ion transfer reactions given by Frahn (1980 [9.10]). Thus the DWBA radial integrals (9.2.13) are replaced by the 'Sopkovich prescription' of factoring out the square roots of the nuclear parts of the partial-wave elastic S-matrices in the initial and final channels, and the remaining transfer radial integrals with Coulomb wave functions are approximated by their WKB forms. The result is

$$R^{(\kappa')L}_{l'l}(k', k) \approx \frac{2\pi}{\zeta' k k' \kappa'} I^{(\kappa')}_{L-K}(\vartheta, \xi)\}\mathring{S}^{(N)}_{l'}(k')\mathring{S}^{(N)}_{l}(k)\}^{1/2}, \tag{9.5.3a}$$

where ϑ and ξ are defined in the same way as in (9.3.6) and ζ' is a scale factor that partly accounts for recoil effects see Frahn (1980 [9.10]) and Kahana and Baltz (1977 [9.17]). The definition and main properties of the functions $I^{(\kappa)}_{L-K}(\vartheta, \xi)$ are given in Appendix 2. A similar approximation of the radial integral for the reverse transfer process gives

$$R^{(\kappa)L}_{ll'}(k, k') \approx \frac{2\pi}{\zeta k' k\kappa} I^{(\kappa)}_{LK}(\vartheta, -\xi)\{\mathring{S}^{(N)}_{l}(k)\mathring{S}^{(N)}_{l'}(k')\}^{1/2}, \tag{9.5.3b}$$

where $\kappa = (2\mu_C E_B)^{1/2}/\hbar$ is the imaginary wave number for particle c in the initial nucleus A. On using the symmetry relation

$$I^{(\kappa)}_{LK}(\vartheta, -\xi) = I^{(\kappa)}_{L-K}(\vartheta, \xi) \tag{9.5.4}$$

the quantity (9.2.16) becomes

$$T^{(\kappa,\kappa')}_{l'l}(k', k) = \frac{4\pi^2}{\xi\xi'(kk')^2\kappa\kappa'} I^{(\kappa)}_{L-K}(\vartheta, \xi) I^{(\kappa')}_{L-K}(\vartheta, \xi). \tag{9.5.5}$$

The modified S-matrix (9.2.14) has the form

$$S_l(k) = \{1 - t^{(\kappa,\kappa')}_{l}(k', k)\}\mathring{S}_l(k) \tag{9.5.6}$$

with

$$t^{(\kappa,\kappa')}_{l}(k', k) = \sum_{K=-L}^{L} \alpha^{(T)}_{LK}(\bar{\lambda}) I^{(\kappa)}_{L-K}(\vartheta, \xi) I^{(\kappa')}_{L-K}(\vartheta, \xi), \tag{9.5.7}$$

where we have written

$$\alpha^{(T)}_{LK}(\bar{\lambda}) = \frac{2\mu^2}{\hbar^4} \frac{a^{(\kappa)}_{T}}{\zeta k\kappa} \frac{a^{(\kappa')}_{T}}{\zeta' k' \kappa'} \tag{9.5.8}$$

(since the coefficients a_T in general depend on L and K and are slowly varying functions of $\bar{\lambda}$). We note particularly that the contribution to the elastic S-matrix (eqn (9.5.6)) from transfer channel coupling is, apart from the factor $\mathring{S}_l(k)$, completely determined by the characteristics of the transfer process. In particular, the relative phase is fixed, and since the function $t^{(\kappa,\kappa')}_{l}$ is mainly real, the transfer coupling contribution is predominantly *absorptive*.

9.6 One-step and two-step elastic transfer processes

One-step elastic transfer is a well-known contribution to elastic scattering of nuclei of the type A = (B + c), so that the transfer of particle c leads back to the elastic channel B + A (Temmer 1962 [9.18]; von Oertzen 1970 [9.19]) as shown schematically in Fig. 9.2. The corresponding effective interaction $\tilde{U}_l(r)$ in eqn (9.2.6) is a real *exchange* potential of the form

$$\tilde{U}_l(r) = (-)^l a^{(\kappa)}_{ET} F^{(\kappa)}_{0}(r). \tag{9.6.1}$$

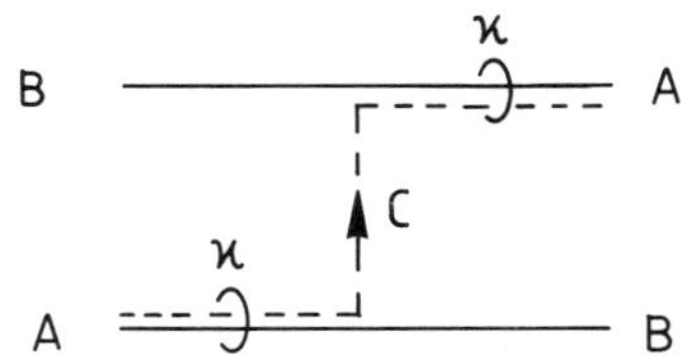

FIG. 9.2. Schematic illustration of the symmetric transfer reaction $A+B \equiv (A-C)+C+B \rightarrow (A-C)+(B+C) \equiv B+A$.

A simple modification of the formalism in Section 9.2 then leads to the following form of the total S-matrix:

$$S_l(k) = \{1 - i(-)^l \alpha_{ET} I_{00}^{(\kappa)}(\vartheta, 0)\} \mathring{S}_l(k), \tag{9.6.2}$$

where

$$\alpha_{ET} = \frac{\mu}{\hbar^2} \frac{a_{ET}^{(\kappa)}}{\zeta k \kappa}. \tag{9.6.3}$$

Thus the contribution of one-step elastic transfer is predominantly *refractive* and gives rise to an 'odd–even staggering' in the partial-wave amplitudes. (An expression similar to (9.6.2) was derived in Section 10 of Frahn (1975 [9.6]).

A two-step (sequential) elastic transfer is a possible contribution to elastic scattering of nuclei of the type $A = (B+c+c)$, for which the intermediate channel is symmetric with nuclei $A' = (A-c) = B' = (B+c)$. This is shown schematically in Fig. 9.3(b). An example is a two-step α transfer in $^{16}O + ^{24}Mg$ with the symmetric intermediate channel $^{20}Ne + ^{20}Ne$. This reaction has been studied recently by Landowne and Wolter (1979 [9.20]) by means of a full coupled-channel calculation (with recoil and finite-range interaction but neglecting non-orthogonality contributions) including coupling to the inelastic channel with ^{24}Mg in its lowest 2^+ state. Although the two-step contributions to the elastic cross-section were found to be at least three orders of magnitude smaller than the direct scattering cross-section at 90°, an *ad hoc* increase of the two-step amplitude by a factor of 5 would be able to account for the weak oscillatory structure observed in the elastic $^{16}O + ^{24}Mg$ scattering at $E_{lab} = 29$ and 33 MeV (Carter *et al.* 1976 [9.21]).

The contribution of the two-step transfer process depicted in Fig. 9.3(b) corresponds again to an exchange interaction

$$\tilde{U}_l(r) = (-)^l \tilde{U}_l^{(T)}(r), \tag{9.6.4}$$

where $\tilde{U}_l^{(T)}(r)$ is the effective interaction corresponding to the process shown in Fig. 9.3(a). From the formalism of Section 9.2, our results

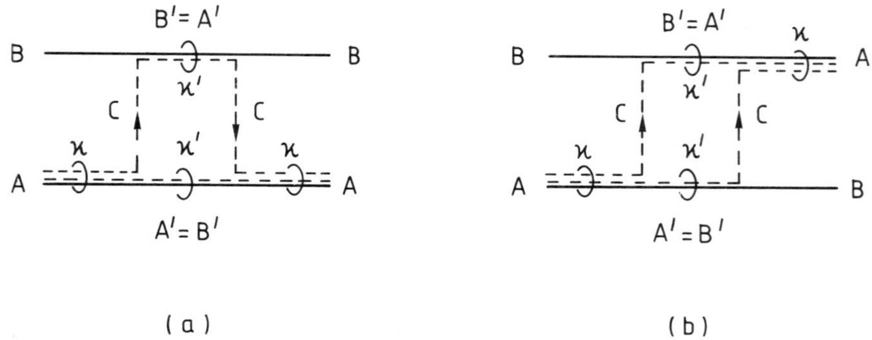

FIG. 9.3. Schematic illustration of possible types of two-step symmetric transfer reactions.

(9.5.6)–(9.5.8), and inspection of Fig. 9.3, it follows immediately that the S-matrix for direct elastic scattering plus the sum of the contributions from the processes of Fig. 9.3 becomes

$$S_l(k) = [1 - \{1 + (-)^l\} t_l^{(\kappa,\kappa')}(k', k)] \mathring{S}_l(k). \tag{9.6.5}$$

Thus the odd-l contributions vanish while the even-l amplitudes are twice as large as those given by eqn (9.5.7). As in eqns (9.5.6) and (9.5.7) the two-step transfer contributions are predominantly absorptive.

9.7 References

[9.1] Love, W. G., Terasawa, T., and Satchler, G. R. (1977). *Phys. Rev. Lett.* **39,** 66.
[9.2] Love, W. G., Terasawa, T., and Satchler, G. R. (1977). *Nucl. Phys. A* **291,** 183.
[9.3] Baltz, A. J., Kauffmann, S. K., Glendenning, N. K., and Pruess, K. (1978). *Phys. Rev. Lett.* **40,** 20.
[9.4] Donangelo, R., Canto, L. F., and Hussein, M. S. (1979). *Nucl. Phys. A* **320,** 422.
[9.5] Frahn, W. E. and Hussein, M. S. (1980). *Phys. Lett. B* **90,** 358.
[9.6] Frahn, W. E. (1975). In *Heavy-ion, high-spin states and nuclear structure,* Vol. 1, p. 215. International Atomic Energy Agency, Vienna.
[9.7] Frahn, W. E. and Gross, D. H. E. (1976). *Ann. Phys.* (N.Y.) **101,** 520.
[9.8] Frahn, W. E. (1976). *Nucl. Phys. A* **272,** 413.
[9.9] Frahn, W. E. and Rehm, K. E. (1978). *Phys. Rep.* **37C,** 1.
[9.10] Frahn, W. E. (1980). *Phys. Rev. C* **21,** 1870.
[9.11] Frahn, W. E. and Hussein, M. S. (1980). *Nucl. Phys. A* **346,** 237.
[9.12] Frahn, W. E. and Hill, T. F. (1978). *Z. Phys. A* **285,** 315.
[9.13] Thorn, C. E., LeVine, M. J., Kolata, J. J., Flaum, C., Bond, P. D., and Sens, J. C. (1977). *Phys. Rev. Lett.* **38,** 384.

[9.14] Alder, K., Bohr, A., Huus, T., Mottelson, B., and Winther, A. (1956). *Rev. mod. Phys.* **28,** 432.
[9.15] Baltz, A. J., Glendenning, N. K., Kauffmann, S. K., and Pruess, K. (1979). *Nucl. Phys. A* **327,** 221.
[9.16] Doll, P., Bini, M. Hendrie, D. L., Kauffmann, S. K., Mahoney, J., Menchaca-Roca, A., Scott, D. K., Symons, T. J. M., van Bibber, K., Viyogi, M. P., Wieman, H., and Baltz, A. J. (1978). *Phys. Lett. B* **76,** 566.
[9.17] Kahana, S. and Baltz, A. J. (1977). *Adv. nucl. Phys.* **9,** 1.
[9.18] Temmer, G. M. (1962). *Phys. Lett.* **1,** 10.
[9.19] Von Oertzen, W. (1970). *Nucl. Phys. A* **148,** 529.
[9.20] Landowne, S. and Wolter, H. H. (1979). *Nucl. Phys. A* **323,** 161.
[9.21] Carter, J., Clarkson, R. G., Hnizdo, V., Keddy, R. J., Mingay, D. W., Osterfeld, F., and Sellschop, J. P. F. (1976). *Nucl. Phys. A* **273,** 523.

APPENDICES

1 Determination of the mean S-matrix

Here we address the question of how to replace the product of square roots of the S-matrices in the initial and final channels by a single 'mean' S-matrix. Assuming that $S_l(k) = \eta_l(k)\exp\{i2\delta_l(k)\}$ can be interpolated by a function of the continuous variable $\lambda = l + \frac{1}{2}$, with the reflection function $\eta_l(k) \to \eta(\lambda) = \hat{\eta}(\mu)$ characterized by the parameters Λ and Δ in the combination $\mu = (\lambda - \Lambda)/\Delta$, we require

$$\{S_{l_f}(k_f)S_{l_i}(k_i)\}^{1/2} \to \{S^{(f)}(\lambda_f)S^{(i)}(\lambda_i)\}^{1/2} \triangleq S^{(K)}(\bar{\lambda}) = \eta_K(\bar{\lambda})\exp\{i2\delta_K(\bar{\lambda})\}, \tag{A1.1}$$

where $\bar{\lambda} = \frac{1}{2}(\lambda_i + \lambda_f)$ and $K = \lambda_i - \lambda_f$. Then

$$2\delta_K(\bar{\lambda}) = \delta^{(i)}(\lambda_i) + \delta^{(f)}(\lambda_f) = \delta^{(i)}(\bar{\lambda} + \tfrac{1}{2}K) + \delta^{(f)}(\bar{\lambda} - \tfrac{1}{2}K), \tag{A1.2}$$

and our requirement for the reflection functions is

$$\{\eta^{(f)}(\lambda_f)\eta^{(i)}(\lambda_i)\}^{1/2} = \left\{\hat{\eta}\left(\frac{\lambda_f - \Lambda_f}{\Delta_f}\right)\hat{\eta}\left(\frac{\lambda_i - \Lambda_i}{\Delta_i}\right)\right\}^{1/2}$$
$$\triangleq \eta_K(\bar{\lambda}) = \hat{\eta}\left(\frac{\bar{\lambda} - \Lambda_K}{\Delta}\right), \tag{A1.3}$$

with Λ_K and Δ determined as functions of Λ_i, Λ_f, Δ_i, and Δ_f such that (A1.3) is optimally satisfied for functions $\hat{\eta}(\mu)$ of general 'strong absorption profile'.

This is of course trivial in the 'sharp-cut-off' limit where $\hat{\eta}(\mu)$ is a unit step function; then

$$\Lambda_K = \bar{\Lambda} + \tfrac{1}{2}|K - K_0| \qquad \text{(sharp-cut-off)}, \tag{A1.4}$$

where $\bar{\Lambda} = \frac{1}{2}(\Lambda_i + \Lambda_f)$ and $K_0 = \Lambda_i - \Lambda_f$.

For smooth profiles we first solve the problem by choosing a particular functional form of $\hat{\eta}(\mu)$ for which (A1.3) can be satisfied exactly, and then use the result in the general case. If $\Delta_i = \Delta_f = \Delta$ this can be done with Kauffmann's function (Kaufmann 1977 [A1])

$$\hat{\eta}(\mu) = \exp(-e^{-\mu}). \tag{A1.5}$$

Then

$$\frac{1}{2}\left\{\exp\left(-\frac{\lambda_{\mathrm{f}}-\Lambda_{\mathrm{f}}}{\Delta}\right)+\exp\left(-\frac{\lambda_{\mathrm{i}}-\Lambda_{\mathrm{i}}}{\Delta}\right)\right\}=\exp\left(-\frac{\bar{\lambda}-\Lambda_K}{\Delta}\right) \tag{A1.6}$$

yields

$$\Lambda_K=\bar{\Lambda}+\Delta\ \ln\left\{\cosh\left(\frac{K-K_0}{2\Delta}\right)\right\}. \tag{A1.7}$$

For $\Delta_{\mathrm{i}} \neq \Delta_{\mathrm{f}}$ we obtain

$$\frac{1}{2}\left\{\exp\left(\frac{K-K_0}{2\Delta_{\mathrm{f}}}\right)\exp\left(-\frac{\bar{\lambda}-\bar{\Lambda}}{\Delta_{\mathrm{f}}}\right)+\exp\left(-\frac{K-K_0}{2\Delta_{\mathrm{i}}}\right)\exp\left(-\frac{\bar{\lambda}-\bar{\Lambda}}{\Delta_{\mathrm{i}}}\right)\right\}$$
$$\hat{=}\exp\left(-\frac{\bar{\lambda}-\bar{\Lambda}}{\Delta}\right), \tag{A1.8}$$

which can no longer be satisfied exactly. However, we can determine 'optical' values of Λ_K and Δ by the following requirements. First, eqn (A1.8) should hold exactly at $\bar{\lambda}=\Lambda_K$. This yields

$$\Lambda_K=\bar{\Lambda}+\Delta\ \ln\left[\frac{1}{2}\left\{\exp\left(\frac{K-K_0}{2\Delta_{\mathrm{f}}}\right)+\exp\left(-\frac{K-K_0}{2\Delta_{\mathrm{i}}}\right)\right\}\right]. \tag{A1.9}$$

Then we determine Δ such that in the case of 'K-matching', i.e. $K=K_0$, where (A1.8) becomes

$$\frac{1}{2}\left[\exp\left\{\left(\frac{1}{\Delta}-\frac{1}{\Delta_{\mathrm{f}}}\right)(\bar{\lambda}-\bar{\Lambda})\right\}+\exp\left\{\left(\frac{1}{\Delta}-\frac{1}{\Delta_{\mathrm{i}}}\right)(\bar{\lambda}-\bar{\Lambda})\right\}\right]\hat{=}1, \tag{A1.10}$$

it minimizes the difference between the exponentials:

$$\left(\frac{1}{\Delta}-\frac{1}{\Delta_{\mathrm{f}}}\right)^2+\left(\frac{1}{\Delta}-\frac{1}{\Delta_{\mathrm{i}}}\right)^2=\text{minimum} \tag{A1.11}$$

which gives

$$\Delta=\frac{2\Delta_{\mathrm{i}}\Delta_{\mathrm{f}}}{\Delta_{\mathrm{i}}+\Delta_{\mathrm{f}}}. \tag{A1.12}$$

Clearly, for $\Delta_{\mathrm{i}}=\Delta_{\mathrm{f}}$, (A1.9) reduces to (A1.7), and both expressions reduce to (A1.4) in the sharp-cut-off limit.

An example of how relation (A1.3) with (A1.9) and (A1.12) works for reflection functions with a 'Fermi function' profile, i.e. $\hat{\eta}(\mu)=(1+\mathrm{e}^{-\mu})^{-1}$, is shown in Fig. A1.

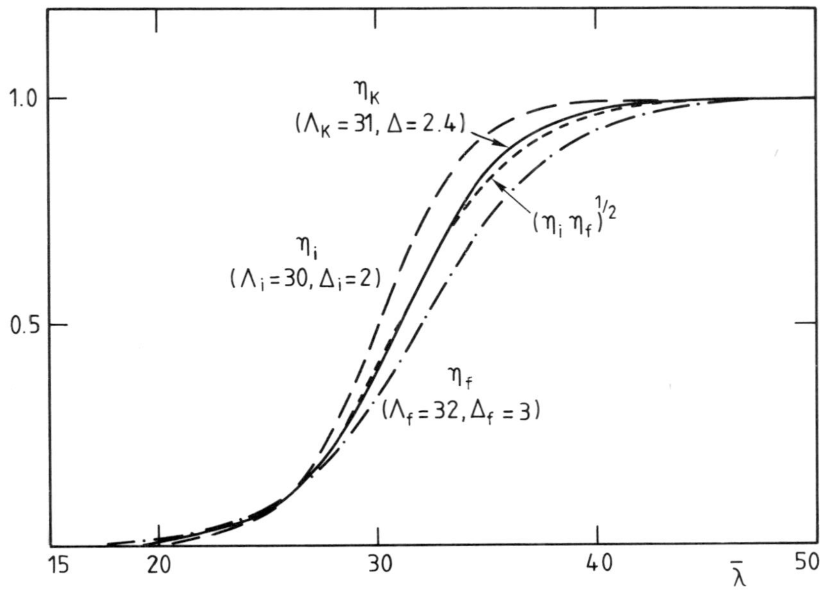

FIG. A1. Mean reflection function $\eta_K(\bar{\lambda})$ of a Fermi function profile (——) calculated from the initial-channel reflection function $\eta_i(\bar{\lambda})$ with $\Lambda_i = 30$ and $\Delta_i = 2$ (— — —) and the final-channel reflection function $\eta_f(\bar{\lambda})$ with $\Lambda_f = 32$ and $\Delta_f = 3$ (—·—). $\eta_K(\bar{\lambda})$ represents the 'optimum' symmetric approximation (eqn (A1.3)) to the asymmetric function $\{\eta_i(\bar{\lambda})\eta_f(\bar{\lambda})\}^{1/2}$ (- - -) with parameters $\Lambda_K = 31$ and $\Delta = 2.4$ calculated from eqns (A1.9) and (A1.12) respectively.

2 The functions $I^{(\kappa)}_{LM}(\alpha, \xi)$

The functions $I^{(\kappa)}_{LM}(\alpha, \xi)$, which are defined in eqn (8.2.22) as the WKB approximation to the Coulomb radial integrals (8.2.14) for the transfer form factor (8.2.5), have the integral representation

$$I^{(\kappa)}_{LM}(\alpha, \xi) = \int_0^\infty dw(-i^L)h_L^{(1)}\{i\rho(1+\varepsilon \cosh w)\}\rho(1+\varepsilon \cosh w)$$
$$\times \cos\left[\xi'\varepsilon \sinh w + \xi w + M \arctan\left\{\frac{(\varepsilon^2-1)^{1/2}\sinh w}{\varepsilon + \cosh w}\right\}\right], \tag{A2.1}$$

where

$$\varepsilon = (\sin \tfrac{1}{2}\alpha)^{-1} = \{(\bar{\lambda}/n)^2 + 1\}^{1/2}, \tag{A2.2}$$

and ρ, ξ, and ξ' are defined in eqns (8.2.23) and (8.2.32). An approximate expansion, with an accuracy of about 10 per cent for $|\xi'|<1.5$, is given by Broglia and Winther (1972 [A2]) as

$$I_{LM}^{(\kappa)}(\alpha,\xi)=\left(\frac{\pi}{2}\right)^{1/2}\exp\{-\rho(1+\varepsilon)\}\sum_{m=0}^{L}\frac{(L+m)!}{(L-m)!m!}\times\frac{1}{\{2\rho(1+\varepsilon)\}^m}\frac{\exp(-b^2/2a_m)}{a_m^{1/2}}, \tag{A2.3}$$

with

$$a_m=\rho\varepsilon+m\frac{\varepsilon}{1+\varepsilon}, \qquad b=\xi'\varepsilon+\xi+M\left(\frac{\varepsilon-1}{\varepsilon+1}\right)^{1/2}. \tag{A2.4}$$

For $\rho(1+\varepsilon)>1$ and $b\lesssim 1$ the sum in (A2.3) can be approximated by the $m=0$ term

$$I_{LM}^{(\kappa)}(\alpha,\xi)\approx\left(\frac{\pi}{2\rho\varepsilon}\right)^{1/2}\exp\left\{-\frac{b^2}{2\rho\varepsilon}-\rho(1+\varepsilon)\right\}. \tag{A2.5}$$

As discussed by Broglia and Winther (1972 [A2]), this expression shows the Q-value and M-dependence of $I_{LM}^{(\kappa)}$ by the exponential of $b^2/2\rho\varepsilon$. For a given angle α, the optimum Q-value is determined by the adiabaticity parameters, giving a maximum $I_{LM}^{(\kappa)}$ for $b=0$, or

$$\frac{\xi'}{\sin\frac{1}{2}\alpha}+\xi=-M\tan\tfrac{1}{4}(\pi-\alpha), \tag{A2.6}$$

and the steepness with which $I_{LM}^{(\kappa)}$ decreases for increasing $|Q-Q_{\text{opt}}|$ is controlled by the binding parameter κ through $\rho\varepsilon=\kappa n/k\sin\frac{1}{2}\alpha$ such that the fall-off is faster for weakly bound particles than for strongly bound particles. The same relation shows that, for positive Q-values, $I_{LM}^{(\kappa)}$ is largest for $M=-L$ and smallest for $M=L$, and that the difference is more pronounced at smaller angles α than at larger α.

For $L=0$, (A2.1) can be evaluated exactly in terms of a modified Hankel function of imaginary order (Trautmann and Alder 1970 [A3]; Trautmann and Alder 1971 [A4]; Alder *et al.* 1972 [A5])

$$I_{00}^{(\kappa)}(\alpha m\,\xi)=\exp\left\{-\left(\rho+\xi\arctan\frac{\xi'}{\rho}\right)\right\}K_{i\xi}(n\gamma\varepsilon)$$
$$\simeq\left(\frac{\pi}{2n\gamma\varepsilon}\right)^{1/2}\exp\left\{-\left(\rho+\xi\arctan\frac{\xi'}{\rho}+n\gamma\varepsilon\right)\right\}, \tag{A2.7}$$

where $\gamma=(\rho^2+\xi'^2)^{1/2}/n$ with the asymptotic form of (A2.7) obtained for $n\gamma\varepsilon\gg|\xi|$.

The main dependence of $I^{(\kappa)}_{LM}$ on $\bar{\lambda}$ is given by the factor $\exp(-n\gamma\varepsilon)$, which becomes $\exp(-\gamma\bar{\lambda})$ for $\bar{\lambda} \gg h$. This enables us to define

$$I^{(\kappa)}_{LM}(\alpha, \xi) \equiv \tilde{I}^{(\kappa)}_{LM}(\alpha, \xi)\exp(-\gamma\bar{\lambda}), \tag{A2.8}$$

where $\tilde{I}^{(\kappa)}_{LM}(\alpha, \xi)$ varies relatively slowly with $\bar{\lambda}$. Finally, using an amended notation, we note from (A2.1) the symmetry relation

$$I^{(\kappa)}_{L-M}(\alpha, \xi, \xi') = I^{(\kappa)}_{LM}(\alpha, -\xi, -\xi'). \tag{A2.9}$$

References

[A1] Kaufmann, S. K. (1977). *Z. Phys. A* **282,** 163.
[A2] Broglia, R. A. and Winther, Aa. (1972). *Phys. Rep.* **4C,** 153.
[A3] Trautmann, D. and Alder, K. (1970). *Helv. Phys. Acta* **43,** 363.
[A4] Trautmann, D. and Alder, K. (1971). *Ann. Phys. (N.Y.)* **66,** 884.
[A5] Alder, K., Morf, R., Pauli, M., and Trautmann, D. (1972). *Nucl. Phys. A* **191,** 399.

PUBLICATIONS OF W. E. FRAHN

1 Articles in professional journals

1 Nichtlokale Feldtheorie auf der Grundlage der Salpeter–Bethe Gleichung. I. 'Freie' Teilchen. *Z. Naturforsch.* **8a** (1953) 620–8. (With H. L. Jordan.)

2 Nichtlokale Feldtheorie auf der Grundlage der Salpeter–Bethe Gleichung. II. Wechselwirkung mit lokalisiertem Teilchen. *Z. Naturforsch.* **9a** (1954) 572–8. (With H. L. Jordan.)

3 On the nucleon–nucleus interaction. *Nuovo Cim.* **4** (1956) 313–22.

4 Nucleon–nucleus interaction from the statistical model. *Nuovo Cim.* **5** (1957) 393–401.

5 Effective nuclear potentials. *Nuovo Cim.* **5** (1957) 523–5. (With R. H. Lemmer.)

6 Velocity-dependent nuclear interaction. *Nuovo Cim.* **5** (1957) 1564–72. (With R. H. Lemmer.)

7 Non-static effects on individual nucleons in a spheroidal potential. *Nuovo Cim.* **6** (1957) 664–73. (With R. H. Lemmer.)

8 Velocity-dependent nuclear interaction. II. Level schemes. *Nuovo Cim.* **6** (1957) 1221–4. (With R. H. Lemmer.)

9 Beugung elektromagnetischer Wellen in Braunbekscher Näherung. I. *Z. Phys.* **156** (1959) 78–98.

10 Beugung elektromagnetischer Wellen in Braunbekscher Näherung. II. *Z. Phys.* **156** (1959) 99–116.

11 Scattering of high-energy nucleons by a nonlocal potential. *Phys. Rev.* **118** (1960) 269–70. (With R. H. Lemmer and Y. C. Tang.)

12 The electromagnetic isotope separator in Pretoria. *Nucl. Instrum. Methods* **7** (1960) 253–68. (With W. L. Rautenbach and L. Wahlin.)

13 Complex potential model for low-energy neutron scattering. *Ann. Phys. (N.Y.)* **16** (1961) 387–406. (With H. Fiedeldey.)

14 P-wave neutron strength function. *Ann. Phys. (N.Y.)* **19** (1962) 428–47. (With H. Fiedeldey.)

15 Surface absorption of low-energy neutrons. *Nucl. Phys.* **38** (1962) 686–95. (With H. Fiedeldey.)

16 Strong absorption model for elastic nuclear scattering. Part I. *Ann. Phys. (N.Y.)* **24** (1963) 243–88. (With R. H. Venter.)

17 Strong absorption model for elastic nuclear scattering and polarization of spin-$\frac{1}{2}$ particles. Part I. *Ann. Phys. (N.Y.)* **27** (1964) 135–66. (With R. H. Venter.)

18 Strong absorption model for elastic nuclear scattering and polarization of spin-$\frac{1}{2}$ particles. Part II. *Ann. Phys.* (*N.Y.*) **27** (1964) 385–400. (With R. H. Venter.)

19 Strong absorption model for elastic nuclear scattering and polarization of spin-$\frac{1}{2}$ particles. Part III. *Ann. Phys.* (*N.Y.*) **27** (1964) 401–24. (With R. H. Venter.)

20 Strong absorption model analysis of 15 MeV deuteron elastic scattering. *Nucl. Phys.* **59** (1964) 641–50. (With L. P. C. Jansen.)

21 Diffraction model for single-nucleon transfer reactions between complex nuclei. *Nucl. Phys.* **59** (1964) 651–69. (With R. H. Venter.)

22 Note on approximations to non-local nuclear interaction. *Nucl. Phys.* **66** (1965) 358–60.

23 Analysis of ^{3}He-nucleus scattering and polarization. *Nucl. Phys.* **74** (1965) 65–72. (With G. Wiechers.)

24 Fresnel and Fraunhofer diffraction in nuclear processes. *Nucl. Phys.* **75** (1965) 577–91.

25 Strong absorption model for inelastic nuclear scattering. *Nucl. Phys.* **80** (1966) 434–48. (With J. M. Potgieter.)

26 Analysis of elastic and inelastic nuclear scattering. *Phys. Lett.* **21** (1966) 211–13. (With J. M. Potgieter.)

27 Proton–nucleus interaction at 20 BeV. *Phys. Rev. Lett.* **16** (1966) 810–13. (With G. Wiechers.)

28 Elastic and inelastic nuclear scattering of heavy ions. *Nucl. Phys.* **92** (1967) 84–90. (With J. M. Potgieter.)

29 Partial-wave analysis of 20 BeV proton–nucleus scattering. *Ann. Phys.* (*N.Y.*) **41** (1967) 442–55. (With G. Wiechers.)

30 Relations between proton–nucleus and proton–nucleon interaction. *Phys. Lett. B* **24** (1967) 216–18.

31 Analysis of quasi-elastic (^{3}He, t) reactions. *Phys. Lett. B* **26** (1967) 5–7. (With G. Wiechers.)

32 Quasi-elastic scattering to isobaric analogue states. *Nucl. Phys. A* **107** (1968) 129–38.

33 Quasi-inelastic (^{3}He, t) reaction. *Nucl. Phys. A* **113** (1968) 593–7. (With G. Wiechers.)

34 Generalized diffraction model for one- and two-nucleon transfer reactions. *Nucl. Phys. A* **133** (1969) 593–609. (With M. A. Sharaf.)

35 Angular-momentum matching in the strong-absorption model. *Lett. Nuovo Cim.* **1** (1971) 561–4.

36 Diffraction systematics of nuclear and particle scattering. *Phys. Rev. Lett.* **26** (1971) 568–71.

37 Diffraction scattering of charged particles. *Ann. Phys.* (*N.Y.*) **72** (1972) 524–47.

38 Fresnel diffraction in high-energy multiple scattering. *Nucl. Phys. B* **62** (1973) 365–80. (With B. Schürmann.)

39 High-energy approximations to nuclear scattering. *Ann. Phys. (N.Y.)* **84** (1974) 147–64. (With B. Schürmann.)

40 Explicit evaluation of the amplitude for elastic heavy-ion scattering. *Ann. Phys. (N.Y.)* **101** (1976) 520–73. (With D. H. E. Gross.)

41 Closed-form quantal description of inelastic heavy-ion scattering. *Nucl. Phys. A* **272** (1976) 413–44.

42 Magnetic substate population in inelastic heavy-ion scattering. *Lett. Nuovo Cim.* **17** (1976) 339–46. (With K. E. Rehm.)

43 Quantum diffraction. *Riv. Nuovo Cim.* **7** (1977) 499–542.

44 Elastic and inelastic heavy ion scattering. *Phys. Rep.* **37C** (1978) 1–51. (With K. E. Rehm.)

45 Dynamic polarization by Coulomb excitation in the closed formalism for heavy ion scattering. *Z. Phys. A* **285** (1978) 315–22. (With T. F. Hill.)

46 Generalized Fresnel model for very heavy ion scattering (I). *Nucl. Phys. A* **302** (1978) 267–80.

47 Generalized Fresnel model for very heavy ion scattering (II). Relation to the optical model. *Nucl. Phys. A* **302** (1978) 281–300.

48 Generalized Fresnel model for very heavy ion scattering (III). Dynamic polarization effects. *Nucl. Phys. A* **302** (1978) 301–9.

49 Theory of elastic and quasi-elastic heavy-ion collisions. *S. Afr. J. Phys.* **1** (1978) 133–42.

50 Closed-form description of polarization in light and heavy ion scattering: elastic scattering. *Ann. Phys. (N.Y.)* **124** (1980) 1–36. (With T. F. Hill.)

51 Closed-form description of polarization in the inelastic scattering of light and heavy ions. *S. Afr. J. Phys.* **3** (1980) 11–23. (With T. F. Hill.)

52 Closed-form description of heavy-ion transfer reactions based on distorted-waves theory. *Phys. Rev. C* **21** (1980) 1870–82.

53 Large-angle scattering of heavy ions (I). General structure of elastic cross section. *Nucl. Phys. A* **337** (1980) 324–50.

54 Large-angle scattering of heavy ions (II). General structure of inelastic cross section. *Nucl. Phys. A* **337** (1980) 351–64.

55 Effective potentials for heavy-ion scattering derived from channel coupling. *Phys. Lett. B* **90** (1980) 358–63. (With M. S. Hussein.)

56 Coupled-channels extension of the closed formalism for heavy-ion collisions (I). Elastic scattering. *Nucl. Phys. A* **346** (1980) 237–65. (With M. S. Hussein.)

57 Nuclear optics. *S. Afr. J. Sci.* **76** (1980) 368–73.

58 Excitation of the lowest 1^- state in ^{18}O by scattering from ^{16}O. *S. Afr. J. Phys.* **4** (1981) 13–15. (With J. Carter, R. G. Clarkson, V. Hnizdo, F. Osterfeld, A. Richter, and J. P. F. Sellschop.)

59 *S*-matrix description of anomalous large-angle heavy-ion scattering. *Nucl. Phys. A* **369** (1981) 166–80. (With M. S. Hussein, L. F. Canto, and R. Donangelo.)

2 Chapters in books

1 Phenomenological description of nuclear scattering and direct interactions. In *Fundamentals in nuclear theory*, pp. 3–91. International Atomic Energy Agency, Vienna, 1967.

2 Structure information from direct nuclear reactions. In *Theory of nuclear structure*, pp. 297–340. International Atomic Energy Agency, Vienna, 1970.

3 Streuung schwerer Ionen. In *Kernreaktionen mit Schweren Ionen* (eds. R. Lipperheide and U. Wille), Part II, pp. 1–79. Hahn-Meitner-Institut für Kernforschung, Berlin, 1972.

4 Recent advances towards a physical foundation of molecular biology. In *7th Annual Seminar on Theoretical Physics, Johannesburg, 1972*, Vol. II, pp. 1–127.

5 Wave mechanical aspects of heavy ion collisions. In *Lecture notes in physics* Vol. 33, pp. 102–26. Springer, Berlin, 1975.

6 Wave mechanics of heavy ion collisions. In *Heavy-ion, high-spin states and nuclear structure*, Vol. I, pp. 157–253. International Atomic Energy Agency, Vienna, 1975.

7 Elastic heavy-ion scattering. In *Heavy ion science* (ed. D. A. Bromley), Vol. I. Plenum, New York, in the press.

3 Conference Proceedings

1 Polarization in elastic nuclear scattering of strongly absorbed spin-$\frac{1}{2}$ particles. *Proc. 2nd Int. Symp. on Polarization Phenomena of Nucleons*, pp. 235–7. Birkhäuser, Basel, 1966.

2 Nuclear interaction of 20-GeV protons. *Proc. Int. Conf. on Nuclear Physics*, Gatlinburg, TN, p. 178. Academic Press, New York, 1967.

3 Superfluids. *2nd Annu. Seminar on Theoretical Physics, Cape Town, 1967*, pp. 1–30.

4 Direct nuclear reactions. *Proc. Int. School on Theoretical Nuclear Physics, Predeal, Romania, 1969.*

5 Frontiers of nuclear physics. *5th Annu. Seminar on Theoretical Physics, Pretoria, 1970*, pp. A1–16.

6 Analysis of excitation functions for elastic scattering of oxygen nuclei. *6th Annu. Seminar on Theoretical Physics, Cape Town, 1971*, pp. 1–18. (With K. M. Hartmann.)

7 Fresnel approximation to high-energy nuclear scattering. In *High-energy physics and nuclear structure* (*Proc. Int. Conf. on High-Energy Physics and Nuclear Structure, Uppsala, Sweden, June 1973*) (ed. G. Tibell), pp. 184–7. North-Holland, Amsterdam, 1974. (With B. Schürmann.)

8 Wave mechanical description of elastic and inelastic heavy ion scattering. *Proc. European Conf. on Nuclear Physics with Heavy Ions, Caën, September 1976*, p. 8. (With K. E. Rehm.)

9 Generalized Fresnel model for very heavy ion scattering. *Proc. Int. Conf. on Nuclear Structure, Tokyo, September 1977*, p. 576. Published by the Organising committee of Conference.

10 Polarization in elastic and inelastic scattering of heavy and light ions. *Proc. Int. Conf. on Nuclear Structure, Tokyo, September 1977, p.* 577. Published by the Organising committee of Conference. (With T. F. Hill.)

11 Coupled channel effects in peripheral reactions in the closed formalism. *Proc. Int. Symp. on Heavy Ion Physics from 10 to 200 MeV/amu, Brookhaven National Laboratory, July 1979*, Vol. 2, pp. 779–90. (With M. S. Hussein.)

4 Reports and Lecture Notes

1 Elementarteilchen. *Jahrb. Rheinisch-Westfälischen Technischen Hochschule Aachen* **6** (1954) 118–28.

2 Antinukleone. *Tydskr. Wet. Kuns* (1957) 140–56.

3 Algebra and physics. *Math. Colloq.* (*U.C.T.*) **2** (1965) 92–6.

4 Strong absorption models for nuclear reactions and hadron collisions at high energies. *Rep. SMR 3/41*. International Centre for Theoretical Physics, Trieste, 1967.

5 The gospel of analyticity. *Math. Colloq.* (*U.C.T.*) **3** (1967) 111–18.

6 Occupation number representation. *Rep. SMR 6/1*. International Centre for Theoretical Physics, Trieste, 1969.

7 Quantum statistical physics. *Lecture Notes*, University of Cape Town, 1971.

8 Research in physics at UCT. *Impulse* **2** (1972) 1–12.

9 What is science? *Impulse* **4** (1977) 6–7.

10 Individuality and identity—'unfamiliar' aspects of quantal reality (Heisenberg Memorial Lecture, University of the Witwatersrand, Johannesburg, November 4, 1976). *Rep. NPRU 77/2*, pp. 1–25. Nuclear Physics Research Unit Report, University of the Witwatersrand, Johannesburg.

5 Works in preparation for publication

1 Analytic evaluation of the eikonal amplitude for strong-absorption scattering. Preprint IPNO/TH 80-44, University of Paris, Orsay, September 1980.

2 A simple representation of nonlocal interactions in collision processes. Preprint IPNO/TH 80-48, University of Paris, Orsay, October 1980.

3 Partial coherence in heavy-ion reactions. Preprint, Hahn–Meitner Institute for Nuclear Research, Berlin, August 1981. (With K. M. Hartmann and W. Dünnweber.)

SUBJECT INDEX

AUTHOR INDEX